GIS FOR BIOLOGISTS
A Practical Introduction
For Undergraduates

About The Author: Dr Colin D. MacLeod graduated from University of Glasgow with an honours degree in Zoology in 1994. He then spent a number of years outside of the official academic environment, working as, amongst other things, a professional juggler and magician to fund a research project conducting the first ever study of habitat preferences in a member of the genus *Mesoplodon*, a group of whales about which almost nothing was known at the time. He obtained a masters degree in marine and fisheries science from the University of Aberdeen in 1998 and completed a Ph.D. on the ecology of North Atlantic beaked whale species in 2005, using techniques ranging from habitat modelling to stable isotope analysis. Since then he has primarily spent time working as either a teaching or research fellow at the University of Aberdeen. He has taught Geographic Information Systems (GIS) at the University of Aberdeen, the University of Bangor (as a guest lecturer) and elsewhere. He has been at the forefront of the use of habitat and species distribution modelling as a tool for studying and conserving cetaceans and other marine organisms. He has co-authored over 40 scientific papers on subjects as diverse as beaked whales, skuas, bats, lynx, climate change and testes mass allometry, many of which required the use of GIS. In 2011, he created *Pictish Beast Publications* to publish a series of books, such as this one, introducing life scientists to key practical skills and *GIS In Ecology* to provide training and advice on the use of GIS in marine biology and terrestrial ecology.

Cover Image: The oak woodland study area used to study the breeding success of hole-nesting birds, such as redstarts (left) and great tits (right), at the University of Glasgow's SCENE field station on the shores of Loch Lomond in Scotland (see exercise two in chapter twelve for more information). © Ross Macleod.

GIS FOR BIOLOGISTS
A Practical Introduction
For Undergraduates

Colin D. MacLeod

Pictish Beast
Publications

ISBN – 978-1-909832-17-6

Published by Pictish Beast Publications, Glasgow, UK.

Published in the United Kingdom

First Printed: 2015.

Trademarks

Warning And Disclaimer

'Space, the final frontier...'
James T. Kirk,
Captain, USS *Enterprise* NCC 1701

This book is dedicated to those who wish to explore
all aspects of how space and spatial relationships influence
marine organisms and ecosystems, but don't know where to start.

Table of Contents

Preface

GIS, or geographic information systems, are becoming an increasingly important part of modern biological sciences, and this will continue as more and more biologists acquire good GIS skills. However, despite this growing importance, there is a distinct lack of books and resources aimed specifically at undergraduate biologists to help them learn how to effectively use GIS in their studies. Instead, biology students, and other novice biological GIS users, often have to struggle with materials written for geographers, and somehow learn to adapt the skills they develop for their careers as biologists. Understandably, this causes problems for many biologist and leads to them either failing to grasp the important role that GIS can play in so many aspects of biological research, or worse, being put off using this important tool altogether.

This book aims to provide biologists with a practical introduction to using GIS using language which they are already familiar with, and using biological examples to explain how GIS works and why it can be useful to biologists. In addition, it provides practical exercises based on the types of things a biologist would want to be able to do with GIS, which the novice biological GIS user can work through to help set them on the road to using GIS in their future careers in a useful and meaningful way. Thus, this book is not a detailed '*How To …*' guide for using GIS, but instead it aims to act as a primer to give biologists an entry point into the use of GIS in biological research.

This book was developed from the more detailed, and area specific, GIS books (such as *An Introduction To Using GIS in Marine Biology*) which are also available from *Pictish Beast Publications*, and that are primarily aimed at postgraduate and practicing researchers. This means that while it contains some of the same background information, this is provided in a simplified manner in this book. In addition, the information is presented here in the same easy-to-understand format that is used in the more detailed books, especially the instructions for the practical exercises. This means that if readers of this book become sufficiently interested in the use of GIS in biological research, and wish to expand their knowledge of GIS, they can move from the basic introduction contained in this volume to the more detailed books without having to familiarise themselves with a different way of presenting this information.

This book is divided into two sections. The first section provides background information on GIS and its uses in biological research, while the second provides six practical exercises which novice biological GIS users can work through to develop their GIS skills. With one exception (exercise five in chapter sixteen), these practical exercises are based on real biological data sets, and I would like to thank those who provided me with data from their research projects for use in this book. For those who are interested, and where appropriate, specific references, including links to organisational websites, are provided in the introductory section of each exercise. In contrast with the other exercises, exercise five is based on a fictitious disease that was inspired by a post-apocalyptic survival novel called *For Those In Peril On The Sea* by Colin M. Drysdale. However, while this disease is fictitious, it is presented in this book in a biological context that will be familiar to many epidemiologists, and seeks to answer a real epidemiological question about how we can seek to understand how diseases spread in space and time, and what this means for our attempts to control and contain them.

Two versions of the instructions for these exercises are provided, one based around the leading commercially available GIS software package (ESRI's ArcGIS© software) and one based around the leading open source, and so freely available, GIS software package (QGIS or Quantum GIS). This means that students can use the same resource regardless of whether their institution has access to commercial GIS software or not. In addition, it helps students see how the same things can be done in different GIS software packages and prepares them for situations in their future careers where they might not have access to commercial GIS software.

This book would not have been possible without the many students and researchers whom I have worked with over the years and who have inspired me to learn much more about GIS in order to help them with their research than I ever would have done on my own. So, in no particular order (and with apologies to those whom I may have forgotten), thanks to Graham Pierce, Karen Hall, Jennifer Learmonth, Caroline Schweder, Laura Mandleberg, Barry Nicholls, Sarah Canning, Sonia Mendes, Caroline Weir, Wezddy del Torro, Ruth Fernandez, Ross Macleod and Lee Hastie. Thanks also to Kenny Dale, Helen Rossall, Ann Mayo, Andrea Airns, Melvi Macleod and Melanie Findlay for their comments on early drafts of this book. Finally, and most importantly, thanks to Sarah Bannon for her help and support, both while writing this book, and with life beyond it (as, believe it or not, there is life outside of GIS).

SECTION ONE:

BACKGROUND INFORMATION ON GIS

--- Chapter One ---

Introduction

What Is GIS?

A geographical information system (known as a GIS for short) is a database which allows the user to explore spatial relationships within and between data sets. It generally consists of four basic parts. These are the data itself, the software used to explore the spatial relationships, the hardware used to run the software and store the data, and the users. For most biological research, the only parts which you ever really need to think about are the data and the software as you are the user and you will probably be using a standard laptop or desktop computer to run your GIS software and to store your data.

What software and data you use will depend on what you wish to do. Therefore, it is always good to have a clear aim of what you wish to achieve before you create a GIS project (you are free to change your mind later, but be aware that this might mean making a new GIS if this new aim is not compatible with your original one for some reason). Aims for GIS projects can range from being relatively simple (*I wish to make a map showing my sampling locations*) to being highly complex (*I wish to model the influence of climate change on the distribution of marine organisms*). As a result, different GIS projects may have very different software and data requirements. However, all of them will share the same basic principles and concepts regarding how spatial relationships between different sets of data can be explored. It is these basic principles which this book aims to help you understand and work with.

Throughout this book the term GIS will be used for both a single geographic information system (of the type you would use in your research), and collectively to refer to all geographic information systems. This is because using the more correct term of GISs as the collective term for all geographic information systems is unwieldy and it is confusing to continually switch between the terms GIS and GISs.

Why Is GIS Useful In Biology?

While you might only be interested in learning how to create nice maps and figures for presentations, reports and publications (and there is nothing wrong with that), it is worth remembering that many biology research projects inherently have a spatial component that is worth exploring. This can range from the distribution of sampling sites to survey tracks, capture locations, movements of individual animals and information about the distribution of specific habitat types. As a result, many biological research projects would benefit from the creation of a GIS to explore spatial relationships within and between the data. In particular, while some projects can be done without using a GIS, many will be greatly enhanced by using it (see *www.GISinEcology.com/ casestudies* for some examples of research projects which have used GIS).

The very act of creating a GIS will make you think about the spatial relationships within your data, and will help you formulate hypotheses to test, or suggest new ones to explore. In addition, thinking about your data in a spatial manner will help you identify potential spatial issues and/or biases within it. For example, plotting the spatial distribution of sampling sites may help you see whether sites which are closer to each other are more similar than those which are further apart. This is something which may not be clear if you only look at your data in a simple spreadsheet or database format. If such patterns exist in your data, and are driven by factors other than those you are studying, this may mean that you have something called spatial auto-correlation which violates the assumptions of many statistical techniques. In this case, you will need to deal with this in some way or other. Similarly, by plotting the locations where a specific species of animal has been recorded over different habitat variables, you can start to develop ideas about which of these variables are important for determining the distribution of that species. While this has always been easy to do for things like land elevation, by plotting the locations on a map, by using a GIS you can look at a wider range of variables and often ones which are not as clearly represented on paper maps (such as the direction that a slope faces).

GIS can also be used to make measurements and do calculations which would otherwise be very difficult. For example, a GIS can be used to work out how much of your study area consists of a specific habitat type, or how much of it is over 500 metres above sea level, or has a slope greater than 5°, and so on. Similarly, a GIS can be used to calculate the size of the home range of an individual animal, or the total area occupied by a specific species, or how long your survey tracks are, or how much survey effort was put into different parts of your study area (see exercise four in chapter fourteen). In addition, you can calculate new variables within a GIS. For example, you can use a GIS to calculate slope and aspect of land from information about its elevation (see exercise three in chapter fifteen). Similarly, you can measure the distances between sampling sites, or the most likely path an animal took from one location to another.

GIS can also be used to link data together in the way that is needed for statistical analysis. For example, many statistical packages require all your data to be in a single table, with one line per sample and information about that sample, and the location where it came from, in different columns or fields. A GIS provides you with a way to easily create such tables and populate it with information, such as the elevation at each location, the direction the slope faces and the distance from features such as the nearest river or lake, from other data sets. This makes preparing your data for statistical analysis much simpler.

Finally, while GIS is mostly used for displaying and/or analysing data, a GIS can also provide important information when deciding where and how to collect your data in the first place. Given that working in the field can be complex and expensive, it is important to get your data collection right the first time, and a bit of planning can go a long way to ensuring that your research is successful. There is nothing worse than spending all your time and money collecting lots of data only to find out when you come to analyse it that you are missing a vital bit of information or coverage. For example, you may find that you have not sampled all the available habitats, or that you have not sampled a specific part of your study area properly, or that your sampling locations were too close together or too far apart. While creating a GIS at the planning stages of your fieldwork will not banish such possibilities completely, if done correctly, it will help you reduce, to a minimum, the risk of these issues cropping up at a later date.

In reality, these are just a few of the many things which can be done using GIS and which are useful to biologists, and there are many other things it can do, too. Really, the usefulness of GIS to biologists is only limited by their imaginations, but in order to see the full potential of GIS, biologists first need to learn how to use it properly. This is because without a good basic foundation in GIS, it is unlikely that a biologist will ever to able to use it to its full potential.

Who Is This Book Aimed At?

This book is aimed at undergraduates, and other novice biological GIS users, who wish to gain a basic foundation in the use of GIS in biological sciences. As a result, it aims to keep things simple and to specifically look at them in a practical biological context. This means that this book is probably most useful for those biologists who have never used GIS before, or who have dabbled with it, but been put off because it seemed a bit scary and complicated.

Almost all the currently available training information for using GIS is, to a greater or lesser extent, designed and written for terrestrial geographers. However, the GIS concept has such a wide scope that no two users will apply it in exactly the same way. As a result, learning how to use a GIS in biological research from information written for terrestrial geographers is like trying to learn how to fly a helicopter by reading a training manual for a jumbo jet. Sure, they both fly in a manner consistent with the laws of aerodynamics, but you are unlikely to succeed. This means that in the past, many biologists have been put off using GIS in their research from the very start.

However, a GIS is, in fact, a relatively simple, but hugely powerful, tool. It is simply a matter of learning how to use it to answer the questions you are interested in, and that is unlikely to be achieved by doing tutorials which tell you how to work out the best position for a shopping centre from census information and a road map. Therefore, as well as background information on GIS, this book provides six practical exercises which can be worked through either independently or in the classroom. These practical exercises cover specific practical aspects of using GIS in biological research, and taken together, they provide enough experience with GIS to get the novice biological GIS user up and running. Two sets of instructions are provided for each practical exercise, one set based on the leading commercial GIS software package (ESRI's ArcGIS© software) and one set based around the leading open source, and so freely available, GIS software package (QGIS or Quantum GIS – available from *www.qgis.org*). This means that these exercises can be done by anyone interested in learning about using GIS in biological research, regardless of what GIS software resources are available to them.

--- Chapter Two ---

The Basics Of GIS

The basic concept of GIS is very simple. A Geographic Information System (GIS) is, at heart, a type of database. However, it differs from most normal databases in one key way. Normal databases primarily consist of a series of tables which can be linked together to allow the data within them to be extracted, compared or manipulated based on values in different fields or columns in the individual tables. For example, in a typical type of data set used in biology, you can use a normal database to extract all records of a particular species or to calculate an average mass of individuals in a sampled population.

However, while normal databases are very good at allowing you to manipulate temporal data (meaning that you can select just those data from a particular time period), they have great difficulty in manipulating data in a spatial context. For example, while you can link data on species occurrence to temperature for the month in which it was recorded (if you have such data available), it is not as easy to link it to temperature data from a similar location in space because a normal database cannot easily work out which temperature data point is closest to the location where a species was recorded.

In contrast, a GIS, as well as consisting of a series of tables, also contains information on the spatial distribution of the data. As a result, in a GIS, it is much easier to link data within the tables based on their spatial relationships. For example, the nearest temperature data point to any point where a species was recorded can be identified and linked to it. This means that a GIS can be used to join data together based on their spatial relationships within and between data sets in ways not possible with other types of databases (figure 1). It is this power to compare and manipulate data based on their spatial relationships which makes GIS such a powerful tool for biological research.

The information on spatial distribution is held in a GIS as a series of data layers (analogous to individual tables within a normal database). These data layers not only have the attributes for each record in them, they also have information which defines the areas of the Earth which they represent, and the size, shape and position of any features within it. Based on this information, GIS software can work out how features in different data layers relate to each other. For example, a GIS could be used to work out the values for different habitat or environmental variables, such as water depth or temperature, where a particular species was seen, as indicated by the arrows in figure 1.

As a result, while tables are the key component of normal databases, data layers are the key component of a GIS. Each data layer will represent a specific source of data in a specific way. For example, one data layer may contain information on the locations where a particular species was recorded during a survey, which might be represented by a series of points, while another may contain information about the route of the survey itself, represented as a line. Others might then contain information about the environment, such as water depth, sea surface temperature and seabed sediment type represented in a variety of other ways (figure 1).

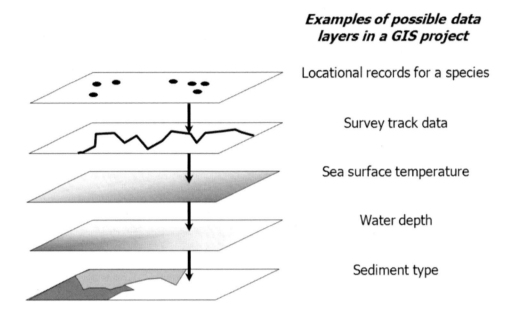

Examples of possible data layers in a GIS project

Locational records for a species

Survey track data

Sea surface temperature

Water depth

Sediment type

Figure 1. *An illustration of how the real world is represented in a GIS project as a series of data layers, each of which contains information about a specific real world characteristic. This could include information about the environment, such as water depth, information about the distribution of organisms in it, such as locational records for a species, or information about human activities, such as the track of a survey through the local area. Due to the way that the data layers are 'stacked' or overlaid on top of each other, the information in different layers can be joined together based on their spatial relationships (as indicated by the black arrows) and not just based on common values in a shared field of a table as is the case in normal databases.*

By adding data layers that contain the specific information you are interested in, to your GIS, you can start investigating the spatial relationships between them. For example, if you wish to know what water depths different species are found in, you can add one data layer which contains information on the locations where each species were recorded, and another which contains information on water depth. You can then compare the spatial relationships of these two sources of data to look at which species occur in which water depths.

A GIS is generally created using specialist GIS software, and such software usually provides a series of tools which allow you to not only create, manipulate and edit data layers (see exercises two and three in chapters thirteen and fourteen respectively), but also to investigate the spatial relationships between them in a variety of ways (see exercise four in chapter fifteen). Therefore, the GIS software that you use is a key component of any GIS project. However, different GIS software may contain different tools, and some are better at some tasks than others. As a result, it is important that, where possible, you choose GIS software which is appropriate to your requirements. More information on GIS software can be found in chapter ten.

--- Chapter Three ---

Common Concepts And Terms In GIS

As with any other technology, GIS is replete with its own language, which at first seems specifically designed to confuse the uninitiated. However, as with any new language, once you know the meaning of a few basic words and phrases, you will be surprised how quickly you can get a basic understanding of it. Therefore, it is worth becoming familiar with what some of the more commonly used terms in GIS mean before you start. If you wish to dive straight in, feel free to skip ahead, but remember this chapter is here in case you get lost. The terms are presented here in an order which allows you to build up a knowledge of GIS terms assuming that you currently know little or nothing about the subject. If you are looking for a specific term, you may find the more extensive, and alphabetical, listing of GIS terms in the glossary in Appendix III (on page 317) more useful.

Projection: A projection is a way of displaying the surface of a three dimensional object (in this case planet Earth) on a two dimensional flat surface, such as a piece of paper (in the case of a traditional map or chart) or a computer screen (in the case of GIS software). As you might imagine, this is not a simple process and always leads to some distortion. As a result, there are many different types of projections, each of which distorts the Earth's surface in a different way to make it fit onto a flat surface. In general, for GIS projects in biology, the most appropriate projection is likely to be either a local, regional or nation-specific projection, such as the British National Grid projection. If one is not available, a transverse mercator (for study areas <~500km in radius) or Lambert azimuthal equal area (for study areas <~1,000km in radius) centred on, or close to, the middle of the study area will usually suffice. However, once you move a long way away from the centre of such projections, the relationships between points start to become distorted. As a result, for larger study areas (such as whole regions or continents), other projections may be more appropriate depending on exactly what you wish to do with your GIS project. For example, you may need a different projection to accurately measure large areas than to accurately measure long distances. In addition, different projections may be required for long (>1,000km), narrow study areas than for square ones in order to minimise the distortion along its entire length.

A brief video explaining what projections are can be found at *www.GISinEcology.com/GFB.htm#video3*, while advice about picking the right projection from those available is covered in more detail in chapter four.

At this point, it is worth mentioning the so called geographic 'projection'. This is not a true projection and it uses decimal degrees as its unit of measurement. This means that it can be used to plot the entire globe. However, a decimal degree does not represent the same distance in all parts of the world. For example, a decimal degree of latitude is shorter, in real terms, at the equator than at the poles due to the curvature of the Earth. In addition, a decimal degree of longitude does not necessarily equal a decimal degree in latitude in terms of real distances. As a result, the geographic projection should not be used for measuring distances or doing calculations based on distances (such as calculating slope). Projections are one of the most important things to understand in GIS and using the wrong projection or trying to compare data which use different projections without taking this into account are some of the most common mistakes made when using GIS.

Coordinate System: Coordinates are the address for any individual point in a GIS. Typically, these are in the form of an X coordinate, or easting, which tells you how far left or right a point is from a given starting point or origin, and a Y coordinate, or northing, which tells you how far up or down it is from this same point. The units (known as map units) for the coordinate system are usually defined by the projection it is associated with. For example, for a transverse mercator projection, the coordinates are usually given in metres, while for the geographic projection, the units are decimal degrees (representing longitude and latitude). When written down, the value for the Y axis (the northing or latitude) may sometimes be given first, followed by the X coordinate (the easting or longitude). Further considerations of the issue of coordinate systems can be found in chapter four.

Spheroid: Planet Earth is not a true sphere. Instead, it is more like a sphere that has been squashed at the ends and has bumpy bits (mountains) and troughs (ocean basins and trenches) all over it. Therefore, in order to produce an accurate map, this slightly non-spherical shape has to be accounted for. This is done by producing a model called a spheroid. Different projections and coordinate systems use different spheroids, which are defined by the datum which is associated with it. For the most part, this is not something you need to worry about except to make sure that you use the same datum for all data layers in a GIS project. If you do not, you may find that they do not overlay each other properly.

Datum: A datum is the starting point on a specific spheroid from which all other points are measured and, therefore, which defines the starting point for the coordinate system. As you can imagine, if different data sets use different datums, then the same point will be plotted in different places (which is not good). As a result, you need to know what datum your data set is based on. Luckily, almost all modern coordinate systems now use the WGS 1984 datum. This is the one used by Global Positioning System (GPS) satellites and most remote sensing satellites. However, if you are using a GPS receiver, these can be set to different datums, so check that you know how yours is set. Similarly, older paper maps, charts and data sets may use different datums. For any one coordinate system, using the wrong datum can result in an error of up to 1km or more. As a result, this is not so important for broad scale and coarse resolution studies, but it is critical for local scale and fine resolution ones. Further considerations of the issue of datums can be found in chapter four.

Map Units: The map units are the units in which the coordinates are provided for a coordinate system. This means these are the units in which distances and areas are measured for a specific projection. In many cases, these will be metres (as is the case with a transverse mercator projection), but this is not always the case. For instance, in the geographic projection, the map unit is decimal degrees. This is important because only when the map unit is in a real measure of distance (such as metres) can distances, areas and other measures be properly calculated. For example, the slope of a hill represents the change in elevation between two points divided by the distance between the points. As a result, slope cannot be calculated when the map unit is not a true measure of distance. This is a common problem with using the geographic projection, and this is the reason its use should be avoided in most biological GIS projects.

Coordinate Reference System (CRS): The coordinate reference system, or CRS, is an alternative name for the projection/coordinate system of a GIS project or data layer.

Decimal Degrees: Decimal degrees are a unit of latitude and longitude. They are generally in the format 12.345678, and the number of decimal places in decimal degrees represents the resolution of such data. The figures after the decimal point represent the minutes, or minutes and seconds, of latitude or longitude as a fraction of a degree. There are 60 minutes in a degree and 60 seconds in a minute. In decimal degrees, coordinates which are north of the equator (0 degrees latitude) and east of the Greenwich Meridian (0 degrees longitude) are positive, while those which are south and west of these lines are negative. Latitude and longitude data must be converted into decimal degrees in order to be plotted in a GIS. It is important when doing this conversion that you know whether your original data are in degrees, minutes and seconds (the traditional format) or degrees and decimal minutes (the standard format for most GPS receivers). More information on converting positions into decimal degrees can be found in Appendix II on page 314.

Data Layer: A data layer is a data set which contains a specific set of information in a specific way in a GIS project. For example, it might represent elevation data as contour lines, or sampling sites as points. You cannot have data in more than one format in the same data layer. For example, you cannot have elevation represented as contour lines and points in the same data layer, as you might see on a terrestrial map or nautical chart. More information about the different ways data can be represented in individual data layers can be found in chapter five. Data layers are often stored as shapefiles which consist of a number of different files and/or folders with the same name, but with different suffixes such as .shp and .dbf. All these separate files are needed for a data layer to be used in a GIS project. As a result, if you are copying any data layer files outside of your GIS software, always make sure that you copy all parts of it.

Feature Data or **Feature Data Layer:** Feature data layers are data layers which contain points, lines or polygons (known as features), each of which are defined by a set of X and Y coordinates (also know as vectors) that mark where they should be plotted in a GIS. Feature data layers are usually accompanied by an attribute table, which is essentially a list that contains information about each point, line or polygon within that specific data layer. A feature data layer cannot contain multiple types of features. For example, a feature data layer can contain either points, lines or polygons, but it cannot contain points and lines, lines and polygons, points and polygons, or all three types of feature together. Since feature data layers are essentially a list of coordinates which are grouped to define individual features, they are not necessarily dependent on the projection and coordinate system in which they were created, and they can easily be transformed into different projections and coordinate systems. While the relative positions of points and lines, and the shapes of polygons, may appear to change during such a transformation, they will still represent the same locations on the surface of the Earth. A feature data layer can represent a grid, where each grid cell is a separate polygon, with accompanying information in an attribute table. This is known as a polygon grid data layer to separate it from a raster grid data layer. An example of how a polygon grid data layer can be used in biological research can be found in the exercise outlined in chapter fifteen.

11

Vector Data or **Vector Data Layer:** This is another name for feature data or a feature data layer.

Shapefile: A shapefile is a specific, and widely used, format for feature data layers, such as points, lines or polygons.

Feature: A feature is a single element within a feature data layer (that is, one containing points, lines or polygons). A data layer may consist of a single feature or it may consist of a large number of different features. Each feature is represented by a unique line in the attribute table and represents the smallest selectable element within a data layer. In order to select only part of a feature, it must first be divided in some way into two separate features represented by their own individual lines within the attribute table of the data layer. The type of feature (that is, whether it is a point, a line or a polygon) within a data layer will depend on how the data are represented within a GIS project.

Vertex: A vertex is a point on a line or a polygon feature. Usually (but not always) vertices are the points where a line changes direction, or mark the corners of polygons, and their positions are defined by a set of X and Y coordinates. This set of X and Y coordinates tells the GIS software where a specific feature should be plotted. The path of a line or the shape of a polygon can be changed by changing the position of one, or more, of its vertices. In addition, the path of a line or the shape of a polygon can be changed by adding additional vertices, which can then in turn be moved to a specific set of X and Y coordinates.

Vertices: The plural of vertex.

Attribute: This is information about a single piece of data (or feature) in a feature data layer. Examples of attributes might include information about its position (e.g. its latitude and longitude), what was recorded there or when it was recorded. Generally, these are displayed in a table (the attribute table) connected to a feature data layer in which each attribute represents a single field or column. If lots of single pieces of data have the same attributes, they can be grouped in the same data layer because they have the same fields. This, of course, does not mean that they have to be. For example, even though you have the same information about every sighting of a particular taxonomic group, you might want to have data on the locations where each individual species was recorded in a separate data layer. If, however, individual features, or groups of features, have different attributes it may be difficult to group them into a single data layer because the fields in the attributes table will not match up.

Attribute Table: A table that accompanies a feature data layer, such as those containing points, lines and polygons, which contains information on the attributes of each feature within that data layer.

Field: A column within the attribute table of a data layer containing information about a specific attribute.

Raster Data, Raster Grid or **Raster Data Layer:** A raster data layer is a data layer which is arranged in a grid, or an array, format. However, unlike feature data layers, each cell within this grid is not a separate feature, and raster data layers are not accompanied by an attribute table which lists the information about each individual grid cell. Instead, the data are stored as an array of rows and columns which give the value for each grid cell, along with information on the size of each grid cell (the height and the width), the number of grid cells in each row and column, and the position of the centre of one of the grid cells (usually the lower left hand grid cell). This information allows the GIS software to know where the raster data layer should be plotted. In general, each cell in a raster data layer is square, and, thus, a raster data layer is very much dependent on the projection and coordinate system which it was created in. As a result, it can be difficult to transform a raster data layer into a different projection and coordinate system. Raster data layers are also known as raster grids to separate them from polygon grids.

Data Frame: A data frame represents the section of the Earth which is being covered by a GIS project. It can be thought of as being analogous to an interactive map and it represents the space where all the data layers within a GIS project are grouped in such a way that they can all be displayed on top of each other and can interact with each other. **NOTE:** There may be problems using data layers in the same data frame if they have different projections, coordinate systems or use different datums to each other and/or from that used for the data frame itself. Therefore, it is good practice for novice GIS users to always ensure that all data layers are in the same projection/coordinate system as the data frame they are being used in.

Layout: This is a way to view the data in your GIS project which allows you to create a figure or map based on its contents. In general, a layout will be attached to a specific data frame, and only those elements displayed in that data frame will be shown in it.

Legend: A legend provides information about the way the data within a data layer are displayed in a GIS project. For example, all features within a data layer with the same value for a specific attribute may be displayed in the same colour, while all features with different values for that attribute are displayed in other colours. The legend tells you what each colour represents. Changing the legend only changes the way in which the data are displayed and does not change the contents of a data layer, or anything in its attribute table.

Overlay: When one data layer sits spatially on top of part or all of another data layer they are said to overlay each other. Two data sets must overlay each other if information needs to be extracted from one based on the positions of features in the other. In many cases, when working with multiple raster data layers, it is essential that the raster data layers overlay each other exactly. That is, their extents must be exactly the same and the edges of their cells must also line up with each other. Data layers which use different projections, coordinate systems and/or datums may need to be transformed into the same projection, coordinate system and datum in order to ensure they overlay each other properly.

Join: A join is when two or more data sets are combined based on a common attribute in a shared field of a table. For example, an identification code could be used to join data from different attribute tables if the same identification codes apply to the same feature in each table.

Spatial Join: A spatial join is when two, or more, data sets are joined based on their spatial locations rather than on common attributes in a shared field. For example, the data from an elevation data layer may be joined to a data layer containing the locations where a particular species has been recorded based on their positions. The ability to do spatial joins is one of the main benefits of using a GIS rather than a normal data base to investigate relationships in your data.

X-Y Data: X-Y data are data which are held in a table that also has coordinates which identify where each data point lies in space. Commonly for biological data, these coordinates consist of latitude (the 'Y' data) and longitude (the 'X' data) recorded from a GPS receiver. In addition to the coordinate data, there may also be other information in the table such as an ID number, what was recorded at that position, and other non-spatial information of interest. In general, X-Y data sets are often created in other types of software, such as databases or spreadsheets. They are then imported into the GIS and plotted using the X and Y coordinates to create a point data layer.

Once added, such point data layers can be used like any other data layer. An example of how to do this can be found in exercise two in chapter thirteen.

Locational Data: These are data which refer to an individual point in space, and generally represent a point where something happened (such as a sampling station) or was observed (such as a sighting of a particular species of interest). They are generally recorded as a set of coordinates along with additional non-spatial information such as time, date, what was recorded and who recorded it. As such, they usually represent 'X-Y Data' and can be treated the same as any other X-Y data set. However, on some occasions, locational data will be recorded as 'waypoints' on a GPS receiver and can be entered into a GIS more directly (see chapters nine and thirteen for more information).

Query: A query is a way of selecting part of a data layer based on the values in its attribute table (sometimes referred to as a table query) or its spatial location relative to features in another data layer (sometimes known as a spatial query).

Interpolation: Interpolation allows you to fill in areas between neighbouring data points of known values. For example, interpolation can be used to convert elevation data points into a continuous surface or grid (an example of this can be found in exercise three in chapter fourteen). Interpolation can be done in a number of different ways (such as inverse distance weighting or kriging) and can be based on varying numbers of neighbouring points. Which approach and what number of points are used will vary depending on exactly what you wish to do. For elevation or water depth data, using inverse distance weighting based on three neighbouring points is typically a useful starting point.

Scale: Rather confusingly, scale is used to refer to two different concepts in GIS. Firstly, scale is used to refer to the resolution used in an analysis or a data layer. For example, a fine scale study of habitat preferences of dolphins might use a depth grid with cells that are 500m by 500m, while a coarse scale study of habitat preferences might use a depth grid of 10km by 10km. Secondly, scale is used to refer to the size of the study area. For example, a local scale study might look at habitat preferences of dolphins within a study area that is a few tens of kilometres across, while a broad scale study might look at habitat preferences across hundreds to thousands of kilometres. This book will use resolution for the first meaning, and will only use scale to refer to the size of the study area.

Resolution: The resolution of a data layer has three related meanings, one in relation to line and polygon data layers, one relating to point data layers and another relating to raster data layers. In line and polygon data layers, the resolution is the smallest features or parts of a feature which are, or can be, represented. For example, a line data layer representing elevation contours would have a resolution of 250m if the smallest topographic features represented within it are those which are 250m, or more, across. In a point data layer, the resolution is the shortest distance over which two events occurring in quick succession would be considered separate points and so be represented by separate rows within the attribute table of the data layer. For example, in a survey of whales, it would represent the minimum distance at which two groups of whales, seen in quick succession, would be entered into a point data layer as two different points rather than one point where two groups were recorded. In a raster data layer, resolution represents the size of grid cells within it. A fine resolution grid will have a small grid cell size while a coarse resolution grid will have a large grid cell size. Within any particular study, you may use raster data layers with different resolutions. For example, you might use a fine resolution raster data layer to look at distribution in relation to elevation, but a coarse resolution raster data layer to look at the distribution in relation to climatic variables, such as rainfall or average summer temperature.

Extent: Extent refers to the area covered by an individual data layer or data frame. For raster data layers, it is set by the coordinates which define the corners of the grid, or by the number of rows and columns, and cell size of the grid.

Precision: This is how precisely a value is measured or expressed. For example, elevation expressed to the nearest 100m will be less precise than elevation expressed to the nearest 1m. However, just because a data layer has a high level of precision, this does not necessarily mean that it has a high level of accuracy. For example, elevation may be measured to the nearest metre, giving it a high level of precision. However, if there is a problem with the altimeter being used to measure elevation above sea level, meaning that all measurements are off by 50 metres (for example due to poor calibration), the elevation measurements will not be accurate, no matter how precise they are (see *www.GISinEcology.com/GFB.htm#video3* for an illustration of the difference between precision and accuracy).

Accuracy: This is how close a value in a data set is to the true world value that it is meant to represent. In GIS, there are two areas where accuracy is important. The first is how accurately values within data layers correspond to

16

the real world values they are meant to represent. For example, within a raster data layer of land elevation an individual grid cell value may be 200m. f the true altitude for that particular location is 205m, then the elevation grid could be said to have a high level of accuracy. If, however, the true altitude is 20m, then the elevation grid could be said to have a low level of accuracy. The second area where accuracy is important is the spatial location of features within a data layer. A high level of spatial accuracy is important to ensure that the spatial relationships between features within a GIS are accurate. In general, you should try to ensure that when you are recording positions of features, such as study sites or locations of species records, you do it with the highest level of accuracy possible. The use of GPS receivers has greatly increased the ability to accurately record locations (down to an accuracy of ~15 metres). However, this does not mean that GPS positions will always be accurate, and problems with receiving signals from multiple satellites can mean that sometimes GPS positions will not be as accurate as they should be, even if they are provided on the GPS receiver in a precise manner (see *www.GISinEcology.com/GFB.htm #video3* for an illustration of the difference between precision and accuracy).

GPS or Global Positioning System: This is a network of satellites which transmit radio signals. When received by a GPS receiver, these signals can be used to calculate the latitude and longitude of the receiver. While this position will also be that of the observer for hand-held GPS units and in small boats, on large ships, the position will represent the position of the receiver's antenna, which may be some distance from where the observer is positioned. While this is generally not an issue, it may be under some circumstances and is something you should be aware of.

Waypoint: A waypoint is a location in space represented by a set of coordinates which is either part of a survey track or which represents a single point in space where an object of interest is located, a point where something happened or a point where something was recorded. The name comes from the use of GPS receivers which can record positions and store them as a list of waypoints.

Digital Elevation Model: A digital elevation model, or DEM for short, is a representation of the topography of the Earth's surface. In biology, this generally refers to the topography of the land or seabed. DEMs can be represented in a number of ways in a GIS, but typically are either a raster grid or a triangulated irregular network (TIN) – see chapter five for more details.

DEM: See Digital Elevation Model.

Continuous Surface: A continuous surface is a data layer which covers the entire area within its extent. Therefore, each point in space is represented by one (and only one) value in the data layer. This may also include a value which indicates when no data are present for a specific location. Continuous surfaces may be non-gridded continuous surfaces, such as triangulated irregular networks (TINs), gridded continuous surfaces, such as raster data layers, or, in the case of polygon grids, feature data layers.

Graticules: Graticules are the markers which provide information about the latitude and longitude for a map. These may be included as numbers around the edges of the map, or as markers on the map itself.

'No Data' Value: A 'no data' value is a value for a cell in a raster data layer which indicates that it contains no information. This is different to a cell containing a value of zero. This is required because all cells within a raster data layer must have a value and cannot be left blank. By including a 'no data' value, GIS software can be told that a specific cell contains no real information and should be treated as if it was empty. Typically, 'no data' values are selected based on the requirements of a specific programme, or on the requirements of a specific data set. The only requirement is that it must not be in the range of real values in the raster data layer. However, in order for a raster data layer to function properly, you may need to tell your GIS software what this value is. If you do not set the 'no data' value, it may not display properly or may cause problems when you use the data layer.

Geodatabase: A geodatabase is a data management and storage framework which is specific to ESRI's ArcGIS software. It allows all data to be stored in a single location and also allows access to some of the more specialist GIS tools which come with that software package. However, for beginners, it is often easier to use what is known as the 'shapefile' approach for storing the data for their GIS projects. In the shapefile approach, which is used for the exercises in this book, each data layer is stored as an individual shapefile within a specific folder (usually on the C-drive of a computer). This allows for easier access to individual files, easier transfer of data layers to other GIS software packages and avoids beginners having to learn about the structure and functioning of geodatabases before they can start using GIS in their research. More experienced users, however, may prefer to use the geodatabase approach to storing data (if their chosen GIS software package allows it). Regardless of the approach used, the same basic rules apply when using GIS for biological research.

Chapter Four

The Importance Of Projections, Coordinate Systems And Datums In GIS

Understanding projections, coordinate systems and datums is paramount to the success of any GIS project. Use them correctly, and your GIS project will be much more likely to go smoothly. Fail to get to grips with them and you may not only encounter difficulties in successfully doing certain tasks, you run the risk of making serious errors in any tasks which you carry out in your GIS project. While these matters may seem complex at first, a bit of careful thought before you start is generally enough to avoid problems with projections, coordinate systems and datums. This is because once you have selected an appropriate projection, coordinate system and datum for a particular GIS project, you will generally not have to change it (although you may have to ensure that all data layers either use these settings or are transformed into them).

Projections:

Planet Earth is a three dimensional place (four dimensional if you consider time as a dimension, but since GIS has problems dealing with time, we will stick to thinking about the world as a three dimensional place for the moment). Maps, including those on your computer screen, by their very nature are flat, two dimensional objects. It can be difficult to accurately represent a three dimensional object on a two dimensional surface. For example, if you cut the skin off an orange, you cannot place it flat on a table without it tearing. In maps, the three dimensional form of the Earth will not necessarily tear as you try to squash it onto a two dimensional map or screen, but it will become distorted. This distortion means that points on the Earth may no longer be in the right places in relation to each other. All two dimensional maps will contain some distortion, the secret is to choose a way of representing the three dimensional form of the Earth on a map so that there is as little distortion as possible for the area and features you are interested in. This is the basis for projections (see *www.GISinEcology.com/GFB.htm#video4* for an illustration of this point).

A projection is a specific way of representing the Earth on a two dimensional map. The term originates from the idea that if you shine a light on the Earth from a specific position, you can 'project' it onto a flat surface, so converting it into two dimensional form. Projections can differ from each other in a number of ways. Firstly, projections can differ from each other in terms of the position of this 'light', and its location in relation to the surface of the Earth. For example, a orthographic projection centred on the North Pole would view the Earth as if it was lit by a light position directly above the North Pole and pointing straight down. In contrast, a gnomonic projection centred on the North Pole would view the Earth as if lit by a light shone from the centre of the Earth itself, while a stereographic projection would view the Earth as if lit by a light shone from a point on its surface directly opposite to the centre of the map (in this case the South Pole as it is opposite the North Pole).

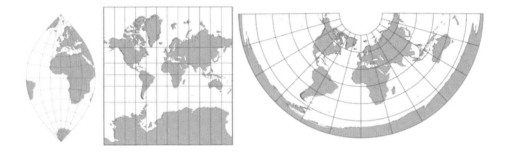

Figure 2. *A comparison of how the world is represented on a flat page using three different types of projection. In all three cases, the maps are centred at 0°N and 0°W. At this point there is no distortion in any of the three projections. The relative positions of all other points are distorted in some way or other. The effects of these distortions can be examined by comparing the shapes of specific land masses, such as Greenland, Africa and South America, in each of the three maps. Left: A transverse mercator projection; Middle: The Mercator projection; Right: An equidistant conic projection.*

Projections can also differ in terms of how they 'unwrap' the spherical Earth. Some projections, such as a transverse mercator projection, will treat it as if it has been projected onto a flat surface which touches the Earth at a specific point and will, more or less, 'unwrap' it equally in all directions as you move from the centre of the projection outwards. The distortion of individual features is proportional to their distance from the centre of the projection, with the greatest distortion at the edges of the projection (figure 2). Such projections are generally only appropriate for relatively small areas. Other projections 'unwrap' the Earth in different ways. For example, the mercator projection (not to be confused with a transverse mercator projection), treats the Earth as if it has been projected onto the inside of a cylinder, rather than a flat sheet, by a light placed at its centre. This cylinder is then 'unwrapped' to show the whole surface of the Earth on a single flat surface. This preserves some characteristics of features over large areas, making them more suitable for larger study areas, but distorts others (figure 2). Other projections may similarly treat the Earth as if it has been projected onto different shapes, such as a cone for conic projections, again with specific effects on the distortion (figure 2).

Any individual projection will usually distort one, or more, of the following three basic characteristics of any feature on a map (see figure 2): the distances between points, the relative bearing or angle between two points and the area of a feature. This distortion may differ latitudinally and/or longitudinally. As a result, it is important to select a projection which will minimise the distortion of whichever of these characteristics you are interested in for your specific area of interest, and whether your study area is relatively square or whether it is long and narrow, and whether its long axis runs north-south or east-west. For example, when navigating at sea, it is the relative bearing between two points which is of most use as this will tell you the direction you need to point your ship in to get from point A to point B, and it is important that charts used for nautical navigation do not distort the bearings between points. However, when analysing data in a GIS, it may be that it is more important to correctly show the distances between points, or the areas of features (as these characteristics may form the basis of your analysis), rather than correctly showing the relative bearings between points.

Coordinate Systems:

Just as a pair of coordinates can be used to plot points on a scatter graph by telling you how far along the X and Y axes each point should be plotted, a pair of coordinates can be used to identify the location of points on the surface of a three dimensional shape such as the Earth. The X coordinates run left to right and are usually referred to as eastings, while the Y coordinates run up and down and are usually referred to as northings. There are a number of different ways this can be done, and different projections use different coordinate systems. These different coordinate systems will be defined by a number of key factors. These are: The position of the centre of the projection (which is equivalent to the origin of a graph and will have a value of zero for both the X and the Y coordinate); the map unit (which is the unit used to measure the X and Y coordinates, and also any distances between two points or areas of features); and the datum (which is the model for the three dimensional shape of the Earth on which it is based).

The most familiar of these coordinate systems is the latitude (the Y variable) and longitude (the X variable) system, which is used for the geographic projection (this is not a true projection, but it will be considered as a projection for the purposes of this discussion). Latitude and longitude measures the position of points on the surface of the Earth in relation to their angle from its centre, and so are measured in degrees, relative to an arbitrary starting point of zero latitude (the equator) and zero longitude (the Greenwich Meridian). Since the coordinates are measured in degrees, this means that the map unit for this coordinate system is a degree, and that any distances between two points, or areas of features, in such a projection will also be measured in degrees. This can cause problems because the actual distance between two points (as measured in real world units of distance, such as metres or kilometres) that are the same number of degrees apart will vary depending on where in the world the two points are positioned. In particular, due to the spherical nature of the Earth, all lines of longitude converge at the poles. This means that the distance between two points which lie on the same latitude, but which are separated by one degree of longitude, will be much closer together at a latitude of $89°$ north than at $1°$ north. Despite this, a projection based on decimal degrees, such as the geographic projection, will treat these two points as being the same distance apart because they are the same distance apart in the map units (degrees), which will be unacceptable in many situations.

Other coordinate systems try to avoid these problems by using real world measurement units, such as metres, as the map units, and define their coordinates in terms of metres from the centre of the projection. However, these coordinates are measured once the surface of the Earth has been projected onto a two dimensional surface. As a result, since projections distort the actual surface of the Earth in order to do this, the relative positions and shapes of features will be distorted, and so will their coordinates. This means that there may still be problems with accurately measuring the distances between points and measuring the area of features. In particular, since these distortions are typically greater the further you go from the centre of the projection (where both X and Y coordinates have a zero value), two points which are the same distance apart in the real world will have different distances when measured in a projection depending on whether they are close to the centre of the projection or far away from it (see figure 3). Specifically, the measurements are likely to be more accurate, in comparison to the real world distances they represent, closer to the centre of a projection than further away from it.

Figure 3. *The effect of changing the position of the centre of a projection (marked by a dark grey dot) on the distortion of features at different locations within a data layer. The distortion is greater the further you get from this central point as illustrated by these three maps which are all in the Lambert azimuthal equal area projection. Left: Centred at 0°N 0°W: Middle: Centred at 56°N 0°W; Right: Centred at 50°S 145°E. Notice that the shapes of Africa, Antarctica and South America are very different between the three images, while some land masses, such as the British Isles, Australia and Greenland can only be plotted in some of these projections due to their positions relative to the central points.*

As a result of this distortion, if you wish to measure distances and areas in a GIS project, it is not only important that you select a projection which does not distort these characteristics, but also that you select one which is centred in an appropriate position so that all the lines and shapes which you wish to measure will be close to this central point. The map units should also represent real world measurement units. Similarly, this means that a projection which may be usable for a local scale area close to the centre of the projection, where the distortion is at a minimum, may not be suitable for a broader scale area, such as a whole continent, as the distortion may become unmanageably large in some parts of the study area.

NOTE: While a specific projection will be centred on a specific point which marks the zero values for the X and Y coordinates of its coordinate system, some also allow you to specify false easting and northing values. These values do not change the point on which a projection is centred. Rather, they change the value of its coordinates. Typically, this is done so that all locations likely to be represented using a specific projection will have positive values. For example, in a projection where the false easting is -5000 and the false northing is -10000, the position of the centre of the projection will be given coordinate values of 5000, 10000 rather than 0, 0.

Datums:

While we usually think of the Earth as a perfect sphere, it is, in fact, far from it. In particular, it is flattened at the poles and bulges around the equator. In addition, it bulges in and out in different places depending on the distribution of things like continents, mountain ranges and ocean trenches. As a result, it is more correctly referred to as a spheroid or an ellipsoid than a sphere. Being a spheroid, it is much harder to build a perfect model of its actual shape, and different projections may use slightly different three dimensional models of the shape of the Earth when working out where points fall on its surface. As a result, a point in the real world may have slightly different coordinates depending on the three dimensional model of the shape of the Earth used for a specific projection and coordinate system. Therefore, it is important that all data sets which you use in a single GIS project are based on the same three dimensional model of the surface of the Earth. The datum is the position of a three dimensional model in relation to the centre of the Earth, and so defines which three dimensional model is used for a specific projection and coordinate system.

Many different datums are currently available and while almost all modern data sources, such as GPS, routinely use the WGS 1984 datum, this should never be assumed to be the case for a specific data layer. In particular, data layers based on older data may be based on different datums. For example, in the UK, data layers based on older Ordinance Survey maps or on the British National Grid projection may be based on the OSGB 1936 datum rather than the WGS 1984 datum. Similarly, while GPS is based on the WGS 1984 datum, individual GPS receivers may be set to display or record positions using a different datum.

How Are Projections, Coordinate Systems And Datums Used In GIS?

While projections, coordinate systems and datums have been treated as separate entities in the above brief introduction to them, in reality they are all connected with each other, and within a GIS, they are generally treated as a single entity. For example, the British National Grid projection uses a transverse mercator projection to 'flatten' the Earth into a two dimensional map, which is centred at 49°N latitude and 2°W longitude. Its map unit is the metre and it is based on the OSGB 1936 datum. In contrast, although the WGS 1984 UTM Zone 30N projection is also a transverse mercator projection and it also uses the metre as its map unit, it uses the WGS 1984 datum and is centred at 0°N and 3°W. Similarly, the geographic projection (which is often the default projection in GIS software and generally the one used for plotting data where the coordinates are in latitude and longitude) uses decimal degrees as its map units and is centred at 0°N and 0°W. It is usually, but not necessarily always, based on the WGS 1984 datum.

As a result, when you select a projection/coordinate system (also known as the coordinate reference system or CRS) within a GIS project for either a data layer or a data frame, you will be selecting the projection, the coordinate system, the position where it is centred, the map units and the datum all in one go. Therefore, while you need to think about all these issues separately, you do not need to set them separately within your GIS.

NOTE: The WGS 1984 UTM Zone 30N datum mentioned above, and other similar UTMs commonly used in GIS, are not true Universal Transverse Mercator (UTM) zone projections, but are based on a simplified version of this system which is centred on the equator. As a result, they may not be suitable for use in GIS projects which are working with data from higher latitudes. Thus, the use of these simplified UTM projections should be avoided in such cases.

What Is The Difference Between Projection/Coordinate System For A Data Frame And For Data Layers In A GIS Project?

In many GIS software packages, including ESRI's ArcGIS software and QGIS, the projection/coordinate system for data layers are independent of the projection/coordinate system of the data frame in which they are being used. As long as the software knows what these are, it can automatically calculate where all the data should plot and, thus, plot them correctly. It can even cope with the inclusion of data layers with different projection/coordinate systems in the same data frame (known as 'on the fly' projection). However, if the software does not know what projection/coordinate system each data layer is in, it will not be able to do this. Therefore, you not only need to set the projection/coordinate system of any data frames in your GIS project, you also need to set the projection/coordinate system of any data layers you are using in it. Luckily, for many existing data layers, this will already have been set. However, whenever you use a data layer for the first time, you should check that this is the case.

While GIS software can usually cope with using multiple projection/ coordinate systems in a single GIS project, for the beginner this can rapidly become very confusing. In particular, some GIS tools will rely on the projection/coordinate system of the data frame, while others rely on those for individual data layers. This can lead to mistakes if certain actions need to be done in a specific projection/coordinate system and different ones are used for the data frame itself and the data layers within it. Therefore, when you first start using GIS, it is best if you try to only work with a single projection/coordinate system in any specific GIS project. This may involve transforming the projection/coordinate systems of some or all of your data layers into the same projection/coordinate system as your data frame before you start using them. Thus, is it good to get into the habit of checking the projection/coordinate system of any data layers when you first add them to your GIS project, and transforming them as required. Instructions on how to transform the projection/coordinate system of a data layer can be found in step two of the exercise in chapter sixteen. In all cases, after doing such a transformation, you need to ensure that it has been done correctly, and that your data layer still plots in the right place. This can be done by comparing the positions where individual features plot in relation to other data layers and to the original, untransformed version of the data layer.

NOTE: There are two very different actions associated with the correct use of projection/coordinate systems for data layers in a GIS. These are assigning the correct projection/coordinate system, and transforming the data layer from one projection/coordinate system to another. Assigning the correct projection/coordinate system to a data layer involves creating an additional file for a data layer which can be used by the GIS software to tell it what projection/coordinate system that data layer was created in. This file usually has the suffix .PRJ and has the same name as the data layer itself. Transforming a data layer involves converting it from this original projection/coordinate system into another one, such as the one you are using for a specific GIS project. You cannot transform a data layer into a different projection/coordinate system simply by changing the definition in the .PRJ file. Similarly, you cannot transform a data layer into a different projection/coordinate system, if its original projection/coordinate system has not been correctly assigned to it.

Key Points To Remember:

This is a very brief introduction to projections, coordinate systems and datums, and if you wish to find out more, then there is plenty of information in other sources. However, it provides you with enough information to start using projections, coordinate systems and datums correctly in your GIS projects. In particular, the main take home points you need to remember from these brief considerations are:

1. All projections will distort features from the Earth's surface in some way, and you should always aim to use a projection which minimises the distortion of the features you are most interested in.

2. The distortions of any projection are generally greater the further from its centre that you travel. Therefore, projections should be centred as close to the centre of your study area as possible. This will define the origin of your coordinate system.

3. Different projections may be required for study areas centred in the same place depending on whether you are working at a local scale or at a broader regional scale due to the different ways they 'unwrap' the planet to show points far away from the centre of the projection. In addition, different projections may be required depending on whether your study area is roughly square or whether it is long and narrow latitudinally or longitudinally.

4. The coordinate system you use should allow you to accurately measure the characteristics of features you are interested in and should have an appropriate map unit to do such measurements.

5. In general, you should ensure that all data layers within a GIS project are based on the same model for the shape of the Earth and so use the same datum. If you do not, you will find that points which should directly overlay each other do not, and this may cause you serious problems. If you have data layers which use projections/coordinate systems based on different datums, you should transform them into projections/coordinate systems which use the same datum before you use them. In addition, it is generally simpler (especially for the novice GIS user) if all data layers in a GIS project, and the data frame, use not only the same datum, but also use the same projection and coordinate system. Again, this may involve having to transform some data layers in order to ensure this is the case.

What Projection, Coordinate System And Datum Should You Use For Your GIS Project?

While it would be great to be able to provide a list of projections/coordinate systems which can be used for different GIS projects, this is simply not possible. This is because the appropriate projection for an individual GIS project will very much depend on what you wish to use it for, the data which you are going to use, and the position and scale of your study area. However, a few general hints may help you identify the most suitable projection/ coordinate system for a specific GIS project. In particular, the following flow diagram takes you through the stages which you need to think about when selecting an appropriate projection, coordinate system and datum for your GIS project.

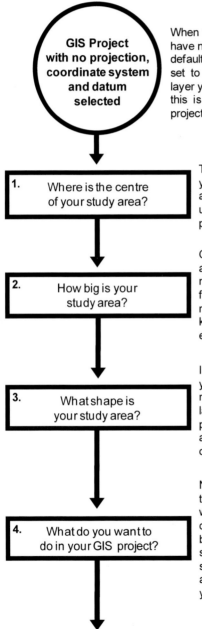

GIS Project with no projection, coordinate system and datum selected

When you start a GIS project, your data frame will either have no projection/coordinate system selected, it will use a default one (such as the geographic projection), or it will be set to the projection/coordinate system of the first data layer you add to it. However, you should always verify that this is appropriate before you do anything in your GIS project.

1. Where is the centre of your study area?

The first thing to consider is what part of the world your study area will cover. This is best represented as the position of its centre as this position can be used to work out what origin is appropriate for your projection/coordinate system.

2. How big is your study area?

Once you have decided on the position of your study area, you need to consider its size. This will help you narrow down the number of appropriate projections for your study area. For small study areas (<~500km radius), a transverse mercator projection of some kind is usually suitable, while for larger areas an equidistant or equal area projection may be required.

3. What shape is your study area?

In addition to considering the size of your study area, you also need to consider its shape, and whether it is relatively square or whether it is long and narrow latitudinally or longitudinally. This is because some projections differ in their level of distortion for latitude and longitude. **NOTE**: This consideration generally only applies to larger study areas.

4. What do you want to do in your GIS project?

Next, you need to consider what you want to be able to do in your GIS project. For example, do you just want to create a map, or do you want to measure distances, or areas, or both, or the relative angles between different points? Different projections will be suited for doing different things, especially for larger study areas. Therefore, it is important that you think about what you wish to do in a GIS project before you decide on a projection/coordinate system.

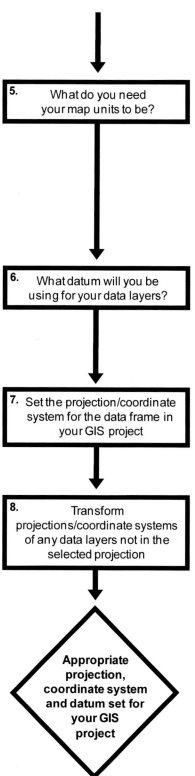

5. What do you need your map units to be?

6. What datum will you be using for your data layers?

7. Set the projection/coordinate system for the data frame in your GIS project

8. Transform projections/coordinate systems of any data layers not in the selected projection

Appropriate projection, coordinate system and datum set for your GIS project

In most cases, you will want to be able to measure characteristics of features in your GIS project in real world measurements, such as metres or kilometres. As a result, you will want the map units of your coordinate system to be in a real world unit. However, on other occasions, you may wish to use different units. For example, the ICES statistical rectangles used in marine biology in the eastern North Atlantic are measured in decimal degrees (and are 0.5° high by 1° wide). Therefore, if you wish to re-create them in a GIS project, you would need to work in a projection/coordinate system which uses decimal degrees as its map unit.

Next, you need to consider the datum which you are going to use for your GIS project. In most cases, the datum will be decided based on the datums of the data layers you want to work with as it is easier to work with data layers using their original datums. In many cases, this will be WGS 1984, but this is not universal, especially for older data sets. If you find that you have data layers which use different datums, you will need to select one, and transform all other data layers into this datum. If you decide to use a different datum from any of your data layers, any which differ will have to be transformed into the new datum.

Once you have selected a projection/coordinate system which is appropriate for your GIS project, it is important to set this as the projection/coordinate system for your data frame before you add any data or do any work in your GIS project.

Finally, you need to transform any data layers which are not in your chosen projection/coordinate system into this projection/coordinate system. This will ensure that you do not inadvertently conduct any work on data which are not in the right projection/ coordinate system, and will ensure that all data layers overlay each other as they should. This may include the data which you recorded on a GPS receiver, as it may have been recorded in a different projection/coordinate system.

For relatively local scale GIS projects, it is often best to use any local, regional or national system which is already in place as these will often have been specifically designed to provide the best representation of features within the individual area they cover. These are generally based on a transverse mercator projection centred within their area of coverage. For example, for GIS projects in the UK, the British National Grid projection may be the most suitable projection. However, not all local, regional or national systems will use the WGS 1984 datum (including the British National Grid projection), and you need to check to ensure that the datum which they use is appropriate for your project and is compatible with any data you wish to include in it.

If there is not a suitable local, regional or national system for your study area, you can usually create your own custom projection/coordinate system centred on the middle of your study area. In many cases, the most suitable starting point for your custom projection/coordinate system will be either a transverse mercator (for study areas <~500km in radius) or Lambert azimuthal equal area (for study areas <~1,000km radius) projection centred on, or close to, the middle of the study area.

For GIS projects which need to cover larger areas, these projections will generally not be suitable due to the distortion which exists as you move further away from the central point and due to the fact that the distortion may be different in different directions. In addition, it may be different for relative bearings, areas and distances. For such projects, other projections are likely to be more appropriate depending on exactly what you wish to do with your GIS project. Similarly, for larger areas, you may find that you need to use different projections to do different things in your GIS project. For example, you may need to use a different projection to calculate areas and distances in large study areas. Information about other possible projections can be found either on the internet or, for ArcGIS 10.3 users, in the online help provided for the software (links to these help page can be found at *tinyurl.com/GFB-Link1*). This information will usually contain advice on the size of area that a projection is suitable for and whether it distorts bearings, distances or areas and to what extent.

For GIS projects which need to cover the entire globe, finding a single suitable projection can be difficult. If you are simply wanting to create a global map with information plotted on it, the mercator projection or the geographic projection will often suffice. However, if you wish to process data or make measurements of individual features, it will be extremely difficult to find an appropriate projection. As a result, you may find that it is easier to break the world down into individual sections which can be analysed separately and for which appropriate projections can be found.

Based on this advice, you can now go back to the flow diagram provided above and customise it for your specific GIS project to aid in selecting an appropriate selection. An example of how to do this is provided below. This example is based on selecting a projection/coordinate system which can be used to calculate the total amount of survey effort which has been conducted in a study area off the northeast of Scotland (see exercise four in chapter fifteen for further details). In this GIS project, you need to add survey data which has been recorded using a GPS receiver set to the WGS 1984 datum, calculate the lengths of individual survey lines and add up all these length values to get the total amount of survey effort in different parts of the study area. However, the study area itself is relatively small (four degrees of latitude by six degrees of longitude). As a result, you will need to select a projection/coordinate system which can be used for this local study area, which uses the WGS 1984 datum and which can be used to accurately measure distances.

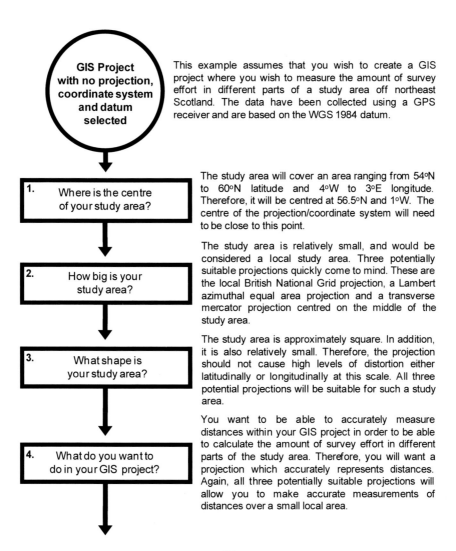

GIS Project with no projection, coordinate system and datum selected

This example assumes that you wish to create a GIS project where you wish to measure the amount of survey effort in different parts of a study area off northeast Scotland. The data have been collected using a GPS receiver and are based on the WGS 1984 datum.

1. Where is the centre of your study area?

The study area will cover an area ranging from 54°N to 60°N latitude and 4°W to 3°E longitude. Therefore, it will be centred at 56.5°N and 1°W. The centre of the projection/coordinate system will need to be close to this point.

2. How big is your study area?

The study area is relatively small, and would be considered a local study area. Three potentially suitable projections quickly come to mind. These are the local British National Grid projection, a Lambert azimuthal equal area projection and a transverse mercator projection centred on the middle of the study area.

3. What shape is your study area?

The study area is approximately square. In addition, it is also relatively small. Therefore, the projection should not cause high levels of distortion either latitudinally or longitudinally at this scale. All three potential projections will be suitable for such a study area.

4. What do you want to do in your GIS project?

You want to be able to accurately measure distances within your GIS project in order to be able to calculate the amount of survey effort in different parts of the study area. Therefore, you will want a projection which accurately represents distances. Again, all three potentially suitable projections will allow you to make accurate measurements of distances over a small local area.

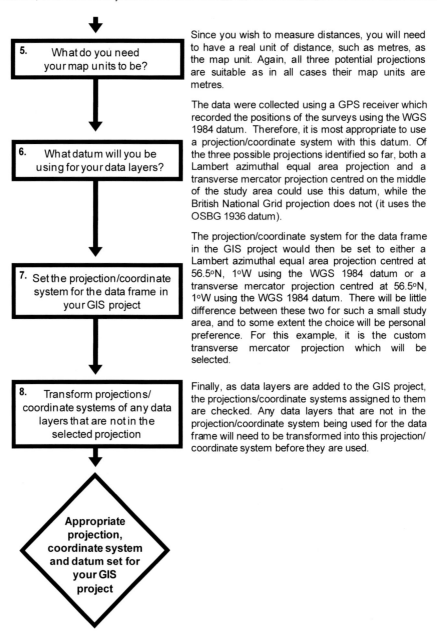

5. What do you need your map units to be?

Since you wish to measure distances, you will need to have a real unit of distance, such as metres, as the map unit. Again, all three potential projections are suitable as in all cases their map units are metres.

6. What datum will you be using for your data layers?

The data were collected using a GPS receiver which recorded the positions of the surveys using the WGS 1984 datum. Therefore, it is most appropriate to use a projection/coordinate system with this datum. Of the three possible projections identified so far, both a Lambert azimuthal equal area projection and a transverse mercator projection centred on the middle of the study area could use this datum, while the British National Grid projection does not (it uses the OSBG 1936 datum).

7. Set the projection/coordinate system for the data frame in your GIS project

The projection/coordinate system for the data frame in the GIS project would then be set to either a Lambert azimuthal equal area projection centred at 56.5°N, 1°W using the WGS 1984 datum or a transverse mercator projection centred at 56.5°N, 1°W using the WGS 1984 datum. There will be little difference between these two for such a small study area, and to some extent the choice will be personal preference. For this example, it is the custom transverse mercator projection which will be selected.

8. Transform projections/coordinate systems of any data layers that are not in the selected projection

Finally, as data layers are added to the GIS project, the projections/coordinate systems assigned to them are checked. Any data layers that are not in the projection/coordinate system being used for the data frame will need to be transformed into this projection/coordinate system before they are used.

Appropriate projection, coordinate system and datum set for your GIS project

Finally, to round off this chapter, it is worth reiterating that whilst understanding projections, coordinate systems and datums is not straight forward, and it can be difficult to work out which is the most appropriate projection/coordinate system for your specific project, taking the time to think about these issues from the start will be repaid in the long run. This is because you will avoid the risk of having to start your whole project again from scratch because you rushed in and only later realised that you did not use the right projection/coordinate system, or that some of your data layers did not use the same datum as others.

Chapter Five

Types Of GIS Data Layers

While some types of spatial data can only be represented in a single way (e.g. discrete sampling locations can only really be represented as points in a point data layer), others can be represented in a variety of different ways. For example, depth information can be represented in a GIS as point data, isobath contour lines, continuous surfaces (such as a TIN), in a grid format (such as a raster grid), or even as polygons which represent areas with the same categories of water depth (figure 4). As a result, care needs to be taken to select the most appropriate way to represent data for the individual task you are conducting. While getting it wrong will not prove fatal to your GIS project, it will waste time if you have to go back and re-create a specific data layer again in a different format.

Examples of different ways of representing the same data

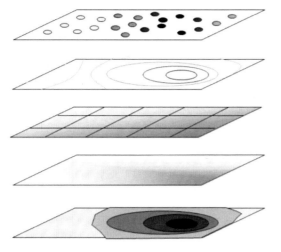

Point depth data
(point data layer)

Depth contour data
(line data layer)

Gridded depth data
(raster data layer)

Non-gridded continuous surface
(TIN data layer)

Polygon depth data
(polygon data layer)

Figure 4. *An illustration of how the same data can be represented as different data layer types within a GIS project. In this case, water depth data could be included as one of five different data layer types, each of which will have its own benefits and limitations for carrying out work in a GIS. Most GIS software will have tools which allow you to convert or transform data between different data layer types (see the exercise in chapter fourteen for an example of this).*

In this chapter, the basic types of data layers which you are likely to come across and use in your GIS projects are described. You do not need to remember all of these as you can always come back and look up what different types of data layers there are, if you need to. However, it is useful to at least skim through them now. Data layer formats can be divided into three basic groups based on how they represent data within a GIS project. These are feature data layers, non-gridded continuous surfaces, and gridded continuous surfaces. These different types can be used in different ways within a GIS project, and often the same variable will need to be represented in a GIS project by different data layer types depending on how you wish to use it.

NOTE: The information in data layers may be saved in a number of separate files, and even folders for some data layer types, with the same name but with difference suffixes, such as .shp and .dbf (see chapter six for more information on this). All these separate files are required for a data layer to work properly. As a result, if you are copying or moving a data layer outside of your GIS software, make sure that you copy or move all the separate files and folders associated with it.

Feature (Or Vector) Data Layers:

Feature data layers, also known as vector data layers, store data as individual features, such as points, lines or polygons. Associated with a feature data layer is an attribute table that holds information about each separate feature in a separate row or line of the table. This means that information about a number of different variables for each feature can be held in the attribute table, with each being stored in a different column or field. There are three basic types of feature data layer: Point data layers, line data layers and polygon data layers. All features in a feature data layer must be of the same type (e.g. all points, all lines or all polygons), and you cannot have a mixture of feature types in a single feature data layer. Features within a feature data layer can overlap, and, indeed, can occupy exactly the same location within your GIS.

Point Data Layer: This is a data layer which contains data that represent discrete point locations. They are typically described by a set of coordinates. Examples of point data layers include ones containing the locations of sampling sites or the positions where individuals of a particular species were recorded. Each unique point will have its own line in the associated attribute table.

Line Data Layer: A line data layer contains information about linear features. Basically, they contain lines that connect locations with the same values for a specific variable. For example, they may connect all locations sampled in a single survey (e.g. the survey route or transect), or they may connect locations of the same altitude (e.g. elevation contour lines). Each unique line will have its own line in the associated attribute table.

Polygon Data Layer: A polygon data layer contains information about an area in space enclosed by a specific boundary (which defines the polygon itself). For example, sampling quadrats could be represented as polygons, as could biogeographic regions, or areas of land. Generally, all locations within a polygon have the same attribute values. Each unique polygon will have its own line in the associated attribute table.

Non-Gridded Continuous Surfaces:

While point, line and polygon data layers can be used to represent discrete features in space, not all data can be neatly divided into such discrete features. For example, the surface of the Earth is a continuous surface with values for characteristics like elevation and slope which continually change as you move across it. Therefore, rather than being represented as discrete features, they can be represented as non-gridded continuous surfaces of some description. The most common of these is a Triangulated Irregular Network or TIN. An alternative to using a non-gridded continuous surface data layer is to use a gridded data layer of some description (see below). However, special care needs to be taken when using a gridded data format to ensure that the resolution of the cells is sufficient to capture all the features within a data set. If this resolution is too coarse, important characteristics of the data set may be left out. If the resolution is too fine, while all the important characteristics will be captured, the file size may become too big and, as a result, the time taken to do anything with it too long. In contrast, in a TIN, the information in the data layer is captured using individual triangles which can vary in size allowing a fine resolution to be used in areas where there is a lot of variation in the data set, and a coarser resolution to be used where there is little or no variation. However, for any specific location, there is only one value for each individual variable represented by a continuous surface.

Gridded Continuous Surfaces:

Gridded data layers can also be used to represent data sets of continuous surfaces. They are defined as a set of regularly-spaced locations, each of which has a value for a specific variable for that location. These locations represent the cells within the grid, and each cell covers a unique and non-overlapping location.

The extent or coverage of a grid is set by both its size and its position in space, and represents the area of the Earth it covers. A broad scale grid will cover a large area, such as an entire continent, while a local scale grid will cover a smaller area, such as an individual mountain, forest or lake. All cells within a grid are the same size, and this size determines the resolution of the grid. For a given extent, a coarse resolution grid will have fewer, larger cells, while a fine resolution data set will have a greater number of smaller cells. The resolution of a grid need not be tied to its spatial extent. This means that a grid with a region-wide extent can have a coarse resolution, if it contains large grid cells, or a fine resolution, if it contains small grid cells. Similarly, a grid with a local scale extent can have a coarse resolution or a fine resolution depending on its cell size.

Gridded data can be represented in a variety of different ways within a GIS, including as feature or vector data layers, each of which has its own advantages and limitations. For some grid formats (such as raster data layers), the grid can generally only be used in the projection/coordinate system in which it was created. As a result, care needs to be taken when creating grids and it is important to know what projection/coordinate system is most appropriate for what you want to do with such data layers before you create them.

Point Grid Data Layers: Grids can be represented as a particular type of point data layer, where each point marks the centre of the grid cell. As a feature or vector data layer, they can be used in ways which differ from raster data layers, which can be beneficial under some circumstances. In addition, point grid data layers are easy to transfer between projections/coordinate systems and between GIS projects. Similarly, since there is an attribute table which has a separate line containing information about each point, point grid data layers provide an easy way of exporting data in a grid for use in non-GIS applications, such as database software, spreadsheet software or statistical software. Finally, point grid data layers can contain information about a number of different variables for each point, with each being stored in separate fields in the attribute table. However, it may not be possible to use point grid data layers in some of the ways which raster data layers can be used (such as in raster calculations – see exercise six in chapter seventeen for an example of raster calculations in action). As a result, other formats are often more appropriate for gridded data.

Polygon Grid Data Layers: A polygon grid data layer is a particular type of polygon data layer where each polygon represents a single cell in a grid. As a feature or vector data layer, they can be used in ways that differ from raster data layers, and which can be beneficial under some circumstances. In addition, polygon grid data layers are easy to transfer between projections/coordinate systems and between GIS projects (although the grid cells may not remain square when transformed into a different projection/coordinate system). Similarly, since there is an attribute table which has a separate line containing information about each polygon, polygon grid data layers provide an easy way of exporting data in a grid for use in non-GIS applications, such as database software, spreadsheet software or statistical software. Finally, polygon grid data layers can contain information about a number of different variables for each polygon with each being stored in a separate field in the attribute table. However, it may not be possible to use polygon grid data layers in some of the ways which raster data layers can be used (such as in raster calculations). As a result, other formats are often more appropriate for gridded data.

Raster Data Layers: In raster data layers, the data are not stored as individual features, but instead are stored as an array of rows and columns, along with information about the size of each cell, the number of cells in each row and column and the position of one of the cells (usually the lower left hand one). They provide an efficient way of storing data about a single variable, and allow you to carry out calculations for all cells in a raster data layer at one time using a raster calculator tool. However, since they do not have an associated attribute table, it is not easy to export such data sets for use in non-GIS applications, such as spreadsheets, databases and statistical software. Raster data layers can be stored in a number of different formats. These include ASCII raster data layers, images, such as JPEGS or GeoTIFFS, CDFs (common data formats) and NetCDFs.

What Types Of Data Layers Should Be Used For Specific Biological Data Sets?

There are no hard and fast rules about what data layer types should be used for specific biological data sets, and in general, you will need to decide how best to represent a specific data set in your GIS. In some cases, the choice will be straight forward as there may only be one way to represent a specific data set in a GIS. However, in most cases, the same data can be represented in a number of different ways. For example, a data layer representing survey effort data could be included in a GIS project as a point data layer to represent specific locations sampled, as a line data layer to represent a transect that was surveyed, as a polygon data layer to represent the survey swath around such a transect, or as a raster data layer where each cell has a value which simply indicates whether it was surveyed or not. Which you choose would depend on how you wish to use the survey data in your GIS project and whether you need to process it further to achieve your intended aim. For example, if you wish to work out the abundance of a species per unit of survey effort (see exercise four in chapter fifteen for an example of this), you would probably choose to represent your survey data as lines. However, if you wish to work out the density of a species per unit of area surveyed, you would probably choose to have your survey data as polygons which represent the survey swath. Similarly, if you wish to simply produce a presence-absence grid of species occurrence, you would probably choose to represent your survey data as a raster data layer.

Therefore, when deciding what type of data layer you need to use for a specific data set, you need to think about the type of data which it represents, how it is currently stored and what your are going to use it for in your GIS project. As with most things in GIS, putting in a little thought in advance can save a great deal of time and effort later. Getting it right from the start will make your life much easier, whilst if you get it wrong you may have to go back and repeat any data processing and/or analysis which you have previously done.

If you find that the data in a specific data layer would be better represented as a different type of data layer, do not panic. In most GIS software packages, it is relatively straight forward to transform the data in a data layer from one type to another. This is particularly useful if you are using existing data layers which have been created by other people and which are provided as a type of data layer which is different from the type that is most suited to the aims of your specific project. For example, data on elevation in a specific study area may be provided as a point or line data layer, and you may decide that you would prefer to have it as a gridded continuous surface, such as a raster data layer. In this case, you would simply use the appropriate conversion tool provided in your GIS software to transform your data layer from a point or a line data layer into a raster data layer (see exercise three in chapter fourteen for an example of this type of transformation).

--- *Chapter Six* ---

Starting A GIS Project

This chapter will outline the information you need, and the thought process you need to go through, to get a GIS project started, and it will allow you to get up and running as quickly as possible. Once you get a bit more familiar with things, you may start deviating from the advice provided here, but this is meant to provide you with a good foundation in terms of how to set up a GIS project so that it will work and be useful to you. This information is provided in a generic format in this chapter, while detailed instructions about how to do specific tasks mentioned here, such as setting a projection and coordinate system for a GIS project, are provided in the practical exercises which can be found in section two of this book.

Things To Consider Before Starting Your GIS Project:

There are a number of key areas which you need to think about before you start your GIS project (and preferably before you start using your GIS software or, even better, before you switch on your computer, so that you are not tempted to dive in too soon):

1. The first thing to establish when starting a GIS project is why you are doing it. If you do not have a clear idea of why you are creating a GIS project from the start, the chances are that you will make mistakes. These mistakes may seem small and insignificant at first, but they have a nasty habit of 'snowballing' and causing bigger problems later if you are not careful. In particular, you might find that you suffer from 'methodological drift', in that as you proceed with your GIS project, your aims slowly change over time, often unconsciously and unnoticed, until you are doing something completely different with your GIS project than when you started out. This can cause problems if the data layers you originally added to your project are not suitable for the new aims which you end up using them for.

For example, you may initially create a GIS project to produce a simple map for inclusion in a report or a publication. For such a map, it may be that you only need a rough approximation of the borders of a study area and the positions of sampling sites within it, and, therefore, when you create data layers to represent these, you may place them in positions which are close enough for the purposes of making an illustrative map. As you start getting more familiar with GIS, you may realise that you can do more with your data layers. You might then come across some existing data layers which show the distribution of different habitat variables in the general vicinity of your study area. You may be tempted to add these data layers to your GIS and start looking at the habitat variables at each sampling site, or the proportion of the study area which contains different habitat types.

This is where problems might arise. Specifically, you did not originally create your GIS to do a habitat analysis of your sampling sites and study area, and your data layers which show these may not be sufficiently accurate to conduct such a habitat analysis (as they were only created to illustrate the approximate positions and not the exact positions needed to make comparisons between different data layers). However, if you do not have a clear idea of your aims in your head when you start your GIS project, you may find that you forget about these limitations and plough on regardless. Having a clear aim of the purpose of your GIS project before you start will not prevent such 'methodological drift', but it will make it much less likely. It will also make you more aware of when you have strayed too far from the original purpose of your GIS project and should, instead, start a new project for your new aim.

2. You need to think about where you are going to store all the data you will use in your GIS project. This may seem a silly thing to think about, but it is very important. In most (but not all) GIS software packages, GIS projects are not like word processing documents where all the information is stored in a single file[1]. Instead, a GIS project is more like a list of addresses which tells the software where all the bits of information are stored. As a result, it is easy to lose track of where all your data are stored on your computer, making it difficult to back it up, and indeed to use the same data in other GIS projects. In general, you should aim to create a single folder on your main storage drive (usually the C:\ or the D:\ drive), and use this to store all your GIS data. This means that you only have to copy a single folder in order to back it up (something which you should get into the habit of doing on a regular basis). Within this GIS folder, you should then create a separate folder for each GIS project and place all the files you will use in it in that single folder, even if this means duplicating data which you use in multiple projects. This is because if you work from the same source file for all GIS projects, it is very easy to edit a data layer in one GIS project only to find that you have deleted something important from it which you needed in a different GIS project, and such deletions are permanent.

---

[1]In some GIS software packages, you can create what is known as a Geodatabase. This represents a single location used to store all the data layers for your GIS project rather than a single file, and each data layer is still stored as a separate entity within it.

3. You need to think about the coverage of your study area. For example, how big an area will it cover and where will it be centred? This is important because it will help you work out the most appropriate projection and coordinate system for your project. It will also help you when you are looking for possible sources of data to include in your GIS project. If you do not have a clear idea of the coverage of your study area for a GIS project, it is all too easy to find that you are using the wrong projection/coordinate system or have ended up using data which do not have the right coverage for your specific aims. While this may not seem like a big problem, if you get it wrong, it can mean having to start again from scratch, which is not only very annoying, but can be very time-consuming.

4. You need to decide what projection/coordinate system is most appropriate to achieve your objectives and for the coverage of your study area. This can seem quite daunting at first as there are many different projection/ coordinate systems to choose from. However, for most biological GIS projects it is simply a matter of using an appropriate local, regional or national projection. If this is not possible, one of two basic options is generally a suitable alternative for GIS projects which will cover a relatively small area in space. These are a Lambert azimuthal equal area projection or a transverse mercator projection centred on the middle of your study area. For larger study areas, different projections may be required (see chapter four for more details). In terms of selecting the most appropriate datum for your coordinate system, most modern data sets, GPS receivers and satellite data use coordinate systems based on the WGS 84 datum, and this is probably best for most projects. The geographic projection (which is often the default projection/coordinate system in GIS software) should generally be avoided as it cannot be used to make accurate measurements, such as calculating the lengths of survey tracks, the distances between data points or the area of features in a data layer. While these recommendations will apply to most projects, and will help you get started quickly, there will be situations where other projection/coordinate systems are more appropriate. Therefore, as you advance your GIS knowledge, it is worth investigating different projection/coordinate systems to see whether they might be more appropriate for any given project.

5. You need to consider where the data that you will use in your GIS project will come from. When doing this, you may wish to consider what data layers you will need, what their resolution is, why they were collected/created, whether this is compatible with the aims of your GIS project, what projection/coordinate system are they in, whether they will need to be transformed into a different projection/coordinate system in order for you to use them, whether you need to create data layers from scratch, or create new data layers from other existing data layers.

Considering such points before you start your GIS project will help you decide on the most appropriate projection/coordinate system for your GIS project and also help define its extent, as well as allowing you to assess whether it is likely to succeed or not.

Things To Consider Before You Add Any Data Layers To Your GIS Project:

Once you have thought about the above issues, you can start your GIS project. However, when doing so, there are a number of additional issues you need to think about::

1. It is generally beneficial to keep a record of what GIS data layers you have or you make, where they are stored on your computer and what information they contain. This can be done using pencil and paper, but it is generally better to create either a database or a spreadsheet where you record all the required information about each of your data layers. This will make it much easier for you to find specific data layers at a later date, and it also allows you to make more effective use of them since you will have all the relevant information you need at your finger tips. This can include information such as the data layer's name, where it is stored on your computer, what data are in it, what projection and coordinate system it is in, when it was created, who created it, what data it was created from, why it was created, the units for any values in it, what projects you have used it in and anything else which you might think would be useful.

2. On a similar note, when you are creating data layers, it is important to ensure that you save them in the right folder on your computer (i.e. the one you set up for the specific project you are using them in) and that you use logical names for them. This is because it is very easy to lose track of data layers if you do not do this. In particular, many GIS tools will use, by default, generic names and may automatically save them in various temporary folders or other odd locations around your computer. If you do not replace these generic names and locations with specific locations and logical names when you create them, you will quickly lose track of what data are stored where. For example, after a while you may find it very difficult to work out what exactly the data layer called 'Depth286' saved in your temp folder actually represents, what project you used it in, and how it differs from the data layers called 'Depth1 to 'Depth285' which are stored in the same folder.

    These problems can be avoided by not only using logical names and saving data layers to the correct folders, but also by creating metafiles for any data layers you create/use. You can store these with the same file name followed by README.TXT (e.g. NW_SCOTLAND_ELEVATION_README.TXT). This file should contain information on: The original data on which it is based, where it came from, why it was created, who created it, when it was created, contact information for that person should anyone have any questions about it, its resolution, its projection and its datum. This not only allows you to identify stray GIS data layers when you come across them on your computer (and this will happen), but it will also make your data layers more useful to others, should you choose to share them.

3. If you are adding an existing data layer into a GIS project, copy the files for that data layer into the correct folder for the specific project before you add them in, and add them in from this new location. If you do not do this first, you will find that you have data layer files stored all over your computer, making it harder to keep track of them and back them up as required.

4. Before you add any existing data layers to your GIS project, you need to ensure that they are suitable for the purpose you are going to use them for. This is generally easy to do, but all too often you will be tempted to rush ahead and use any data layers you come across without thinking about this first. While you may get away with this sometimes, this will cause you problems eventually, so do not get into the habit of using existing data layers without thoroughly checking them out first. When considering whether a data layer is suitable for the purpose you are going to use it for, the things that you need to think about include: What resolution are the data represented by the data layer? Is this resolution suitable for my purposes and coverage? What projection, coordinate system and datum does it use? Will it need to be transformed into a different projection/coordinate system/datum before I can use it? What units does it use for any values in its attribute table? Is it accurate, both overall and for my specific study area? Does it contain the information which I need or should I look for a different source of data? Why was it created and is this purpose compatible with the aims of my GIS project? In many cases, all these questions can be answered by reading any accompanying material for a data layer before you use it. However, as a general rule, if you are in any doubt as to the suitability of a data layer for a specific project, do not use it. There is nothing worse than basing an entire GIS project on a specific data layer only to find out later that it is unreliable or not suitable for your purposes.

Things To Consider After You Have Added A Data Layer To Your GIS Project:

Whenever you add a new data layer to your GIS project, you need to check it both for suitability and for errors. This will help you avoid making mistakes by using unsuitable or erroneous data layers in your project. Much of this can be checked by simply making sure that it plots in the right place relative to other data layers (e.g. no positions which should be on land plot in the sea, or no known major features are missing from elevation data). In addition, simply changing the legend for the data layer allows you to assess whether the values for individual features make sense and are appropriate. For example, within most GIS processes, it is assumed that a zero land elevation value represents sea level. As a result, water depth values are negative, while terrestrial elevations are generally positive. However, many sources which represent purely marine data have depth values which are positive. This can cause problems if such values are processed using standard GIS tools, and such data should always be converted to negative values before you use them. This is an issue which you will only pick up by examining the range of values present within a given elevation data set. Similarly, examining the values for elevation within a data set will allow you to check whether elevation are in feet, metres, or some other unit of measurement.

The Structure Of GIS Projects:

As noted above, in most (but not all) GIS software, GIS projects are not like word processing documents. When you add data layers into them, it is not like pasting a picture into a word processing document. Rather, a GIS project is more like a list of addresses where data are stored on your computer. When you add a data layer to your GIS project, you are simply adding its address to this list so that your GIS software knows to include the data from that specific file in a specific project. This has some important implications. These include:

1. If you wish to transfer a GIS project to a different computer, you cannot simply copy the project file onto a disk and load it onto a second computer. This is because it will only take the list of addresses for all the data layers with it, and will not take any of the data layer files themselves. If you wish to transfer a GIS project to a different computer, you need to not only transfer the project file, but also all the data layer files as well. In addition, these will generally need to be placed in the same location with the same address on the second computer in order for the project file to be able to find them. For this reason, when you start a GIS project, it is generally a

good idea to create a single folder to store all the files used in it. In addition, this file should be stored in such a way that it has a relatively simple address. This makes it easier to put it in the same location on a different computer. For example, if you have all your GIS files for a particular project in a folder with the address C:\GIS_PROJECT, this can easily be move to the same place on a new computer. If, however, you store all the files in a folder on your desktop or in your documents folder, the address will be something like C:\DOCUMENTS AND SETTINGS\USERS\USER NAME\MY DOCUMENTS\GIS_PROJECT. You will not be able to place this in a location on the second computer with the same address unless it has exactly the same pathway available, including the user name, which is unlikely. Saving all your data for a GIS project in a single folder on your main storage drive also allows you to create back ups of these files with relative ease as the whole folder can simply be copied onto a backup disk. This is something you should get into the habit of doing on a regular basis.

2. If you move any of your GIS files to a new location on your computer, any existing projects which use them will not be able to find them. This is because its address within the GIS project file will no longer be correct. As a result, you should avoid moving GIS files around your computer as there is a risk that any GIS projects which include any moved GIS files will stop working. In fact, this is one of the more common reasons why GIS projects stop working properly (see appendix I for more information).

3. One way to quickly and easily move GIS projects between computers is to store them on an external hard drive. However, if you wish to use files stored on an external hard drive, you must ensure that it is provided with the same letter at the start of its 'address' each time it is connected to any computer which you use. On computers running the Windows operating system, this can be done through the Disk Management options.

4. Because a GIS project does not make a copy of any data added to it, but rather relies on the original source file, any changes you make to a data layer in a GIS project will affect the source file. For example, it you delete a feature or edit the contents of the attribute table, these changes will be applied to the actual source file and not just to the version in your GIS project. Therefore, if you delete or change something, it is permanent. In addition, the same changes will appear in any other GIS projects (yours or other peoples) that work from the same source file for that specific data layer. For this reason, whenever you start working with a new data layer, you should always archive a copy of it somewhere on your computer, or on a CD, DVD or external hard drive, so that if you accidentally delete or change something important, you have a copy of the original which you can

go back to and start again. In addition, this means that you should always think twice before deleting anything from a data layer in a GIS project and if in any doubt, make a copy of it before making any permanent changes.

The Structure Of GIS Data Layer Files:

While it is common to refer to data layers as if they are single entities, generally they are not, and there are often several files and/or folders associated with each data layer. All these files and folders are needed for a data layer to work properly in a GIS project. In general, all the files and/or folders will have the same name, but they will have different suffixes. For example, a point data layer stored as a shapefile called SPECIES, may consist of files called: SPECIES.DBF, SPECIES.PRJ, SPECIES.SBN, SPECIES.SBX, SPECIES.SHP and SPECIES.SHX. Each separate file contains different information. For example, SPECIES.DBF will contain the information contained in the attribute table of the data layer, while SPECIES.PRJ contains information about the projection, coordinate system, datum and map units which the data layer was created in.

This has three main implications. First, if any of these files are deleted, the data layer will usually stop functioning. Second, if you want to copy or move a data layer to a new location on your computer outside of your GIS software, you need to copy or move all the files and folders which make it up and not just some of them. Finally, since it is in a widely used database format, it is possible to open and edit the file with the extension '.dbf', which contains the information for the attribute table of a data layer, with other software, such as a spreadsheet programme. This can be much faster than doing it in your GIS software. However, beginners should avoid doing this as it is very easy to accidentally change something which means that the data layer will no longer work in your GIS software.

Sources Of Data For GIS Projects:

Often one limitation of starting a GIS project is finding suitable layers to get you started. Many sources of existing data layers are available and these can be found by conducting searches on the internet, or by asking other people who have done similar projects in the past. However, to help you get started, a list of potential sources of existing data layers for use in GIS projects can be found at *www.gisinecology.com/gis_data_sources.htm* which may be of use and which may allow you to get started with your GIS project as soon as possible.

Chapter Seven

Translating Biological Tasks Into The Language Of GIS

GIS software packages have generally been developed by geographers rather than biologists. As a result, they use a very different language to describe the functions of tools available within them than the one that biologists are already familiar with. In addition, different software packages may use different names for tools which do essentially the same tasks. At first, this can make it very difficult for biologists to work out how to use GIS software to do the tasks that they are interested in doing. For example, a marine biologists may look at their data and think *How can I work out the water depth where I saw each species?* This is clearly a very simple question, and one which is of fundamental importance when studying the distribution of marine species. As this is something that geographers rarely need to do, regardless of how useful biologists would find it, it is unlikely that you will find a tool in your GIS software called *Extract Depth Data To Species Locational Records*. However, this does not mean that you cannot do this task in a GIS. Rather, it means that you just need to translate your task from the 'biological' language which you regularly use in your research into 'GIS-speak', the language of GIS. This is not as difficult as it at first seems (so do not be put off by it), and is relatively easy to learn how to do with a few simple pointers.

The first step is to recognise that there are two parts to this translation. Firstly, you need to break the task down into its most basic steps which can be done by individual tools available in your GIS software. Only then can you do the next step, which is working out which tools you need to use. This is analogous to translating a sentence from one language into another by first breaking it down into individual words or phrases, and then using a dictionary to translate each word into the new language, although, ultimately, it will be more successful than this analogy. Of course, as with any translation task, you need to make sure that you choose the right word from the dictionary to make sure that the translated sentence actually makes sense.

One of the easiest ways to work out how to break a specific biological task down into individual steps is to use a simple flow diagram-based approach to build instruction sets which detail the steps which you need to do to complete that task. This approach has the advantage that it is easy to follow, and it allows you to work out exactly what you need to do, and in what order. It also allows you to record exactly what you did. This is extremely important as it allows you to archive your methods. There may be many months between each time you do a specific task, and without clear records you will be faced with having to learn how to do it anew each and every time, which is both annoying and time-consuming. In addition, such flow diagrams give you something to refer back to when the inevitable tricky question is raised by an over-picky reviewer or examiner as to exactly how you processed your data (with apologies to those involved when I have served this role in the past). Similarly, using this type of common format for each task you need to carry out allows you to share your knowledge with others, and, indeed, for you to learn from others about how to do specific tasks. Finally (and this is not something which you necessarily need to worry about at this stage), it allows an easier transition towards the potential of automating tasks which you need to commonly perform, saving you both time and effort which can be better spent on doing your actual research rather than processing data in your GIS.

Working out which specific GIS tool to use for each individual step can be tricky, and will depend on the GIS software which you are using. However, as examples, instructions of how to do a range of such individual tasks in ESRI's ArcGIS 10.3 software, and the freely available QGIS 2.8.3 package are provided in the six practical exercises which can be found in the second section of this book. In addition, information on how to do specific tasks in GIS can be found using search engines, such as Google, and on a range of GIS forums, such as *gis.stackexchange.com*, and the *GIS In Ecology* forum (*www.GISinEcology.com/GIS_in_Ecology_forum.htm*). Finally, instructions on how to do a wide variety of other tasks can be found in the other books in this series, such as *An Introduction To Using GIS In Marine Biology* and its accompanying workbooks (see *tinyurl.com/GFB-Link2* for more information on these books).

Outline Of The 'Translation' Process:

Breaking down a biological task into its constituent parts is relatively straight-forward, and whilst it may seem very difficult the first time you do it, it quickly gets easier. The basic structure of this translation process is a flow diagram that uses different shapes to identify different elements within it, and numbers to identify the order in which the individual steps need to be done.

Figure 5. *The first step in working out how to complete a specific task in GIS is to identify your goal and place it in a diamond at the bottom of the page.*

The first step in putting such a flow diagram together is to think about the goal which you wish to achieve. This is often best done by restating your task as a sentence which begins with a statement such as: *At the end of this task, I want to have ...* For example, you may wish to know which water depths different species were recorded in and, specifically, to be able to test whether two species occur in significantly different water depths using a statistical test, such as a t-test or Mann Whitney U test. In order to run such tests, you generally need to have your data in a series of columns, one with the species identification, one with the location where it was recorded and one with the water depth at that location. Therefore, you can restate your original biological task as: *At the end of this task, I want to have a table which has the locations where the species of interest were recorded and the depth at those locations.* This is the first element which is filled in on the flow chart and is put in a diamond-shaped box at the bottom of the page (figure 5).

Next, you need to think about what information you need to start with in order to achieve this goal. In general, this will involve identifying what data you need to bring into your GIS project and where it will come from. For example, in order to work out the water depths where two species occur, you will need two separate sets of data: 1. Locational records of where the species were recorded; 2. Information on water depth for the same area. These starting points or starting requirements are added to the top of the flow diagram in circles (figure 6).

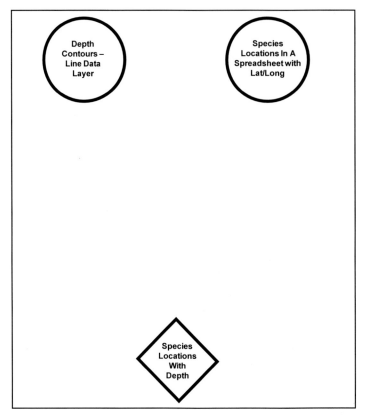

Figure 6. *The second step is to identify what data or information you will need to have before you can start. These are placed in circles at the top of the page.*

The next task is then to join the circles at the top of the page, which represent the starting points or requirements for your task, to the diamond at the bottom of the page which contains the goal. You do not necessarily need to identify all the required data right from the beginning, and you can come back and add more starting points or requirements if you find that you need more of them as you move through working out how to do a task. The starting point circles and the goal diamond are joined by a series of connected rectangles (figure 7). Each rectangle contains information on an individual step you need to do to move from your starting points to your goal. These rectangles are numbered to show which order these steps need to be completed in. Each rectangle contains two parts. The left hand side is filled in with the action of the individual task in

50

plain English, while the right hand side is filled with the name of the appropriate GIS tool, or other processes, required to achieve it. For example, step two in the example being considered here is to plot the species locational data in a GIS (figure 7). In the left hand side of the rectangle for this step is written what is being done in plain English (*Plot species locational data*), while in the right is written the name of the tool in GIS-speak (*Plot X-Y data*).

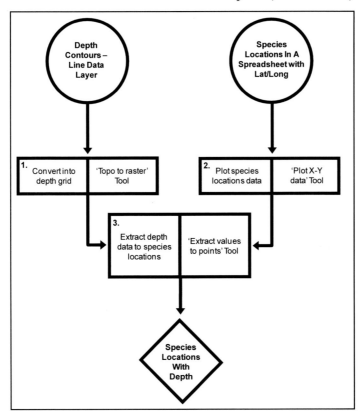

Figure 7. *The final step is to work out and enter the individual steps which you need to complete to get from the starting points contained in the circles at the top of the page to the goal in the diamond at the bottom of the page. These steps are entered as rectangles which are divided into two parts. The left-hand side is used to outline the step in plain English, whilst the right-hand side contains information on what tools are needed or the processes you need to carry out in your GIS to achieve this step. Arrows connect these rectangles to show the flow of work through the process and numbers in each rectangle are used to indicate the order in which the steps must be done.*

By using this flow diagram structure, even relatively complex tasks can be broken down into the individual steps which you will need to carry out in your GIS software to complete them, and they can then be easily followed and repeated each time you need to do a specific task. Do not worry too much about getting it perfect before you start doing a task for the first time. You will often find the best approach is to fill it in as you go along, adding a rectangle each time you identify what the next step you need to do is, and which GIS tool you need to do it. As a result, the best approach for building such flow diagrams is usually to use a software package where you can draw the individual shapes, add the accompanying text and move them around as required. This makes it much easier to amend your diagram if you find that you need more data at the start, that you have missed a step out or, more commonly, that you have to divide what you thought would be one step into two or more in order to be able to do them in your GIS software. In contrast, if you write out your flow diagrams by hand you will find that it can get very messy very quickly, especially if you find you need to change lots of things as you work through the process.

NOTE: Not all steps included in this process are necessarily specific actions in the GIS software itself, and some may include actions which you take outside of your GIS project. For example, you might include a step which involves converting latitude and longitude data into decimal degrees so that you can plot them in your GIS project (see Appendix II), and you might want to do this in a spreadsheet programme before you bring the data into the project itself. In addition, not all steps need to involve actually doing something with your data. Instead, some can represent thought processes or decisions about specific settings you need to use. For example, you might include a step which says something like: *Identify an appropriate projection and coordinate system for measuring distances in my study area.* This process would involve looking on the internet to find out about the properties of various projections and coordinate systems in order to allow you to select an appropriate one. Similarly, you might include a step which involves deciding on the best cell size for a raster data layer, or defining the extent of your study area.

For more complex tasks, you may find that individual flow diagrams become overly complicated. In this case, you can break them down into sections represented by individual instruction sets consisting of flow diagrams with their own starting points and goals. These are then linked together by a summary flow diagram which is used to tell you which order these individual instruction sets need to be carried out in. Using this nested hierarchical approach, tasks that seem very complex at first can be broken down into a series of simple steps which are relatively easy to follow.

This flow diagram approach is the basis of the format used in the practical exercises section of this book. However, in these cases, the information in the right-hand side of the rectangle representing each step has been expanded out to provide explicit instructions on how to carry out each step using a specific GIS software package.

--- Chapter Eight ---

GIS And Statistical Analysis

While most GIS software packages are great for investigating spatial relationships between different data sets, and for linking the information together based on these spatial relationships, in general they are not designed to allow you to conduct the types of statistical analyses required in biological research. This means that in order to carry out statistical analyses, the data will usually have to be exported from your GIS project, analysed in a specific software package and, if required, the results imported back into your GIS project to be plotted again.

In order for you to be able to do this, you will generally need to take the time to work out exactly what information you will need in order to do a specific statistical test and what format it needs to be in. For example, in order to do an analysis of habitat preferences, you will generally need the following sets of information in your GIS: 1. Data on where the species of interest was recorded; 2. Data on other locations which were surveyed or sampled, and where the species was not recorded; 3. Data on appropriate environmental variables, such as elevation, vegetation type and temperature, at those locations. In addition, for most statistical packages, this information will need to be presented in a table format, where each line represents a sampled location, and different fields or columns contain information on species occurrence and environmental information at each of these locations (this is sometimes referred to as the 'big table' format since it effectively contains information from many different sources in a single big table).

Once you have worked out what information you need, and what format it needs to be in, you will need to work out how to link all these data together. This will usually be done by combining all the information into the attribute table of a single feature or vector data layer using a combination of normal joins, spatial joins (for data from other feature data layers) and extraction tools (for data in raster data layers). It is usually easiest to start with a feature data layer which represents the surveyed locations in some way. This can be a point, line or polygon data layer that contains information in the attribute table as to whether each specific location was surveyed. Information can then be added to this attribute table as additional fields which contain information extracted from other data layers, such as whether a species was recorded at a specific location, and how many individuals were recorded, or the values for environmental variables at those specific locations. This process may take some time, and may take several attempts to get right. However, the time spent ensuring that this is done correctly will be well worth it as it will make the statistical analysis much more straight-forward. More information about how information from different data layers can be joined together can be found in exercise four in chapter fifteen.

If you think that you might need to display the results of your statistical analysis in your GIS project, it is important that you have a way to link the results back to your original data layers. For example, you might need to ensure that you include latitude and longitude coordinates in your 'big table' to ensure that it can be re-plotted in your GIS project. Alternatively, you might need to ensure that there is a field in your table which contains a unique ID number of some kind that could be used to join the results of your analysis back into your original data layer for display purposes.

Alternatively, you may find that you can use a field or raster calculator tool to help you get the results of your statistical analysis into your GIS project. For example, if you conduct a linear regression to investigate habitat preferences, you can use the formula generated from your linear regression in one or other of these types of tools in order to visualise how the linear regression model suggests how a species is likely to be distributed throughout a specific study area.

NOTE: When joining data together from different sources, you need to ensure that there is a way to tell which cells in the resulting attribute table contain zero values, and which contain no data (and so should be blank). This can sometimes be difficult as your GIS software may automatically fill in blank cells with zeros. This is because it assumes that no data and a zero value are the same thing – which is often not the case in biology. For example, in a data layer of water temperatures, a value of zero means that it is very cold, where as a blank value means that no water temperature reading was conducted at a specific location. This is an issue which you must keep an eye out for and avoid at all costs. This is because once the cells which should be blank are filled in with zeros, it can be virtually impossible to separate them from the real zero values. The best way to avoid this happening is to make sure that you do not have any missing values, or by using a specific 'no data' code which cannot be confused with any real data values.

--- Chapter Nine ---

Using GPS Data In Your GIS Project

The global positioning system (or GPS) provides a quick and easy way to collect spatial data which can be included in a GIS project. In particular, GPS receivers allow you to record where you have gone and also record locations where you took samples or encountered a specific species of interest. However, there are a number of key points which you need to think about when recording data using a GPS receiver.

Firstly, before you start recording any data with a GPS receiver, you need to check how it is set up. In particular, you need to check what datum it is recording positions in, and what units it is recording the coordinates and any measurements in. In addition, you need to also check what time zone and format it is using to record the time of day, and whether the date has been set properly. This information all needs to be recorded for later use, and the settings changed if any of them are inappropriate. This is because you may find that your data are unusable if you do not have this information. For example, if it is not clear whether your data were recorded in British Summer Time (BST), or Greenwich Mean Time (GMT), you may not be able to work out what stage the tide was at when you recorded a particular species at a particular location. Similarly, if you are not sure whether your coordinates were collected using the OSGB 1936 datum or the WGS 1984 datum, or whether the latitude and longitude were recorded in degrees and decimal minutes or degrees, minutes and seconds, you may find that you cannot plot them in the right places relative to other data. With respect to this, it is important not to assume that all GPS receivers are set up in the same way, or that someone has not changed the settings on a specific receiver since you last used it. As a result, you should get into the habit of checking the settings on any GPS receiver which you use when you first pick it up or turn it on. Information on how to do this is provided in the flow diagram on the next page, and a video version of these instructions can be found at *www.GISinEcology.com/GFB.htm#video9*.

55

Handheld GPS receiver that will be used to collect your field data

For illustrative purposes, the specific steps to set up a Garmin eTrex 10 GPS receiver will be provided. However, the same general steps will apply to all GPS receivers.

1. Open the SET UP menu

On the main page of the GPS receiver, move to and select the SETTINGS option.

2. Set the units and datum that will be used to record positions

Move down to the POSITION FORMAT option and open it. You will now be provided with three options. Select the top option (POSITION FORMAT). You will then be provided with a drop down menu where you can select from a range of possible units to record your position in. The important thing here is to know what units you are using. The most widely used setting in biological research is hddd°mm.mmm'. This records positions in degrees and decimal minutes. However, never assume that all biological data are recorded in this format as some people will use other units.

Once you have selected the units for the position, you need to select the datum that will be used for your positions. This is done in the MAP DATUM section. Select it and you will be provided with a drop down menu. The important thing here is to select the same datum that will be used for your GIS project. The most widely used setting in biological research is WGS 84. However, never assume that all biological data are recorded in this format. Once the datum has been set, you can press the BACK button to go back to the SETTINGS menu.

3. Set the units and the reference for headings

Select HEADING in the SETTINGS menu. In the headings menu, you will find three options. First, select the top option (DISPLAY). This allows you to choose how your headings will be displayed. For biological data collection, you will generallly want to select NUMERIC DEGREES. Once you have selected this, move down to the NORTH REFERENCE option and select it. Here you will find a drop down menu that allows you to set the north reference for your headings. The two most commonly used options are TRUE and MAGNETIC. For most biological purposes, you will want to select TRUE. However, remember that if you are also using a compass when in the field, it will point to magnetic north and you will need to apply a correction to get the two to match up. Once this has been set, click on the BACK button to return to the SETTINGS menu.

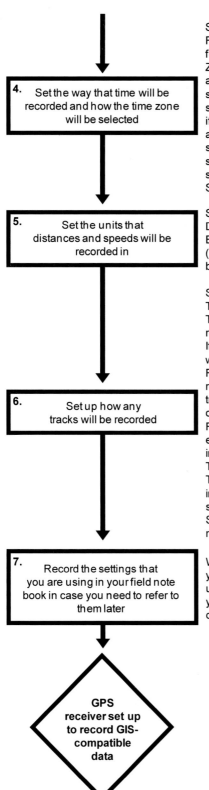

4. Set the way that time will be recorded and how the time zone will be selected

Select TIME in the SETTINGS menu. First, for TIME FORMAT, select 24-HOUR. This is the standard format for scientific data collection. Next for TIME ZONE, select AUTOMATIC. While this will be appropriate in most circumstances, there may be some occasions when you will want to set it to a specific time zone. This will be particularly the case if you are likely to cross from one time zone to another during data collection as you may wish to select one to use for all your data rather than switching between two different ones. Once this is set, click on the BACK button to return to the SETTINGS menu.

5. Set the units that distances and speeds will be recorded in

Select UNITS in the SETTINGS menu. For DISTANCE AND SPEED, select METRIC. Next for ELEVATION (VERTICAL SPEED), select METERS (M/MIN). Once this has been set, click on the BACK button to return to the SETTINGS menu.

6. Set up how any tracks will be recorded

Select TRACKS in the SETTINGS menu. For TRACK LOG, select RECORD, SHOW ON MAP. This means that your track will be automatically recorded and displayed on the GPS receivers map. If you wish to temporarily stop recording a track, you would go to this section and select DO NOT RECORD. However, remember to turn the track recording back on the next time you wish to record a track. Next, select RECORD METHOD. You will be offered three options: AUTO, DISTANCE and TIME. For most biological surveys you will want to select either DISTANCE or TIME. Next, you want to set the intervals that your track data will be recorded at. This is done by selecting RECORDING INTERVAL. This will open a window where you can set the interval for your data collection. Once this has been set, click on the BACK button to return to the SETTINGS menu, and then click BACK again to return to the main menu page.

7. Record the settings that you are using in your field note book in case you need to refer to them later

When you come to use your data in a GIS project, you need to have a record of what settings you used. If you do not have a record of these settings, you may find that you have forgotten them and so cannot use your data in your GIS.

GPS receiver set up to record GIS-compatible data

Secondly, you need to decide how you are going to record your data. While many GPS receivers can be set up to record data automatically, it is important that you check whether the settings are appropriate or if they need to be changed. For example, if you are using a GPS receiver to record your survey track, you need to ensure that it records a position at regular enough intervals to get an accurate representation of where you went, but not so frequently that the memory becomes full before you finish your survey. Similarly, when you are recording the position of events as waypoints, it is important to keep a note of what event each waypoint represents as the name that you use for it may seem logical at the time, but it may be impossible to work out what it refers to at a later date. In addition, you need to ensure that you download your data from your GPS receiver on a regular basis as they have a nasty habit of over-writing older data, especially track data, when the memory gets full.

If you decide to record your data directly on a GPS receiver, you will need to ensure that you have some way of downloading these data from the GPS receiver and converting them into a format which is compatible with your GIS project. This may involve using your GIS software (see step six in exercise two of chapter thirteen), using the standard software available from the makers of the GPS receiver which you use, or using custom software created by a third party. Such custom software can readily be found on the internet, and one of the most useful is the DNRGPS software (available from *www.dnr.state.mn.us/ mis/gis/DNRGPS/DNRGPS.html*) which provides an easy-to-use set of tools for downloading and converting GPS receiver data into a format which is suitable for use in GIS projects (see *www.GISinEcology.com/GFB.htm#video9* for an illustration of how to do this). You will also need the appropriate cable to join your GPS receiver to your computer.

Alternatively, you may choose to record your data in some other way. For example, data may be recorded on a data recording form, or using a digital voice recorder of some kind. If you choose to record data in this way, ensure that you record all the information you need for every occasion. This is relatively easy to do on a data recording form, but may be more difficult to remember to do on a digital voice recorder. In this case, it is worth preparing a 'script' on a small piece of laminated card which is attached to the recorder to ensure that you record all the required information each and every time. In addition, you need to ensure that data are recorded in a comprehensible manner as trying to work out what the numbers in a scrawled latitude and longitude position are meant to be at a later date can prove difficult. Similarly, you will need to ensure that when this information is transcribed into a format which can be loaded into a GIS project (such as a spreadsheet), this is done accurately and that any potential errors are investigated and corrected before you start working with your data (see *tinyurl.com/GFB-Link3* for an outline of how to do this).

Finally, you can often connect your GPS receiver directly to your GIS project through your GIS software. This allows you to add the information from your GPS receiver into your project in real time. This is usually done using a specific tool, such as one of those available on the GPS toolbar in ArcGIS 10.3 or the GPS Information Panel in QGIS 2.8.3.

In addition to downloading data from a GPS receiver into a GIS project, it is also possible to upload information from your GIS project to a GPS receiver. For example, you can use a GIS project to create an outline of your study area, divide it into a grid, mark locations of sampling points or details of a planned survey route. This information can then be converted into a format which can be uploaded to your GPS receiver to aid in your data collection. As with downloading data, this usually involves using specific tools in your GIS software, if they are available, or the use of a specific programme, such as DNRGPS, and requires that you have the appropriate cable to link your GPS to your computer.

NOTE: Most modern smart phones and tablets have build-in GPS receivers, and these can be used to record biological data in place of a dedicated GPS receiver. In order to do this, you will need to install an app which allows you to access the outputs of the GPS sensor. For the Android operating system, one of the most widely used and useful GPS apps is GPS Essentials (available from the *Google Play* store – for more information visit *www.gpsessentials.com*). However, if you are using a smart phone or tablet to record biological data, you need to ensure that it is using the GPS network to work out its position, and not another source of geolocation information. This is because other sources of geolocation information are much less accurate and will not allow you to record your data with sufficient accuracy for use in a biological GIS project. In addition, you need to take similar care to ensure that your GPS app is set up correctly before you start to record your data as you do with a dedicated GPS receiver. Advice and information on how best to do this can be found at *www.GISinEcology.com/GFB.htm#video9*.

--- Chapter Ten ---

An Introduction To GIS Software

A range of software packages are available for creating GIS projects. A limited number of the most popular used by biologists are considered here. In many cases, you may not have a choice in terms of which you use. For example, you may have to go with the software provided by your organisation, or you may have to go with the one which you can afford rather than the one you might ideally wish to use. However, it is always worth knowing what else is on offer, as some programmes are better at some specific tasks than others. Therefore, if you are lucky enough to have access to more than one GIS programme, you may find it easier to use different software packages for different tasks.

In addition, it is worth pointing out at this stage that some tasks may be quicker and easier to do in other types of software. For example, complex non-spatial joins are usually easier in a normal database programme. As a result, if you get stuck trying to do a specific task, try thinking 'Outside-The-Software' for a solution. That is, ask yourself, *Can I get the same result using a different type of software?* as sometimes it may be quicker and easier to do using non-GIS software.

For the purposes of the practical exercises section of this book, details of how to do the tasks are provided for both ArcGIS 10.3 and QGIS 2.8.3 software packages. This is because these are the most widely used programmes in the academic context at the time of writing. In addition, these are the software packages most people use when first learning how to use GIS in their research. However, this does not represent an endorsement or recommendation of these software packages by the author, nor does it mean that the makers of these software packages endorse the contents of this book.

ESRI GIS Software:

ESRI is one of the leading providers of GIS software, and over the years they have produced a range of different products. These products have often become the industry standard for GIS software and are widely used in academic research. Of the products they have produced, the most widely used in biological research have been ArcView 3 and ArcView 9, and most recently ArcGIS 10 (with the latest version, at the time of writing, being 10.3).

ArcView 3 was particularly widely used as it represented an easy-to-use programme that was great for simple tasks, such as quickly putting a map together or just plotting your data to have a quick look at it. There were also a large number of extensions (software add-ons which extend the functionality of the core programme) freely available on the web to do a mind-boggling variety of specific tasks ranging from calculating an animal's home range to working out which route would be the most fun to ride a motorbike along (generally the wiggliest). The chances were if there was something you wished to do which could not be done with ArcView 3 itself, someone would have written an extension to do it. Some of the most useful extensions included Spatial Analyst, X Tools, Animal Movement and Vector Conversion. However, while it was widely used in the past, ArcView 3 is not supported by Windows 7 or later. As a result, it has become much less widely used in biological research.

ArcView 9 and ArcGIS 10 are more recent incarnations of ArcView. These software packages represent a very different interface from ArcView 3, meaning that those who have learned GIS on the earlier version of ArcView may need to put in a substantial amount of effort to transfer their skills to ArcView 9 or ArcGIS 10. However, its much greater functionality more than repays this effort. The main factor limiting the use of ArcView 9 or ArcGIS 10 is the cost of the software licences for the core programme and essential extensions, such as Spatial Analyst. However, reduced cost licences may be available for some users, such as non-profit organisations, under specific circumstances. In addition, relatively cheap home licences are available for non-commercial use. For more information about how to get a licence for ArcGIS 10 for non-commercial uses, visit the *GIS In Ecology* Forum (*tinyurl.com/GFB-Link4*).

IDRISI GIS Software:

IDRISI is a series of software packages from Clark Labs, with the latest incarnation at the time of writing being IDRISI Selva (which has now been incorporated into the TerrSet suite of integrated GIS analysis and image processing tools). It provides much of the same GIS functionality of ESRI products and in some areas may provide greater functionality. In particular, many people find it easier to deal with time series data and data from remote sensing sources within IDRISI software. While the required licences cost less than those for ESRI products and discounts are available for some academic users, they may still prove prohibitive for others.

Manifold:

Manifold represents a GIS solution which is available at a lower cost than many other commercial GIS software packages. It has much of the same functionality of more expensive GIS software packages, however you need to ensure that you obtain the right licences to access all the GIS tools that you might need. In addition, it is based on a very different approach to GIS than most other GIS software. In particular, unlike most other GIS software packages, the GIS project file contains all the data which has been added to it, rather than it simply representing a list of addresses where the files containing the data are stored outside of the project file itself. This means that some people may find it difficult to transfer their GIS skills between the Manifold GIS approach and that used by other GIS software packages. As a result, Manifold is not recommended as a suitable GIS software for novice biological GIS users to develop their GIS skills.

GRASS:

GRASS (Geographic Resources Analysis Support System) is one of the leading open source, and so freely available, GIS software packages. This provides many advantages, but the main one by far is the fact that is does not have any licence costs associated with it. It maintains a high level of functionality, and many of the tools available in commercial GIS software are also available in GRASS. In addition, it can read and access data layers in most standard GIS formats, such as shapefiles. In many respects GRASS is the GIS equivalent of R, which has revolutionised statistical analysis in ecology and marine biology in recent years. However, some users may find it difficult to work with due to the nature of the available interfaces (but see below for one solution to this problem).

QGIS:

QGIS (or Quantum GIS – *www.qgis.org*) has an easy-to-use, graphic user interface which provides a quick and simple approach to carrying out many basic tasks in GIS, such as creating data layers and some of the easier tasks involved in manipulating them. In addition, it allows easy access to the powerful GIS tools available in GRASS, as well as other software packages including SAGA and Python. As with GRASS itself, it is freely available and does not have any associated licence costs. It can also read and access data layers in most standard GIS formats, such as shapefiles. This is probably one of the easiest points of entry into GIS for those who do not have the money to pay for GIS licences, but who still wish to be able to use it in their research. It is for this reason, that this has been selected as one of the software packages used in the practical exercises section of this book.

DIVA-GIS:

DIVA-GIS is another freely available programme which can be used for GIS projects. While it may not have quite as much functionality as other GIS software considered here, it does provide an interface which allows the quick and easy creation of maps for presentations, reports and papers, and the manipulation of data with some relatively simple GIS tools. It can read and access data layers in most standard GIS formats, such as shapefiles. If you are looking for a relatively simple GIS programme for doing relatively simple GIS tasks without costly licence fees, this may be the option for you.

R:

An increasing range of GIS tasks can be done in R statistical software. Again, it is freely available, so does not come with costly licence fees. It has the advantage that many complex statistical analyses can also be conducted within the same software package. However, while this may change in the future, it is still much easier to do many GIS tasks in specialist GIS software. In particular, R is primarily code-based rather than graphic user interface (GUI)-based. This can make it much more difficult for the novice GIS user to learn unless they are already familiar with how to use R. In addition, the lack of a map window (see chapter eleven for more information) can result in mistakes going undetected, and this can result in serious errors in any GIS processing conducted in R.

MapWindow:

MapWindow is another open source GIS software package, which, therefore, is available without having to pay for a licence. While it currently lacks some of the tools which biologists might need to use GIS in their research, it has a very usable interface and is easy to learn. In addition, it provides some of the easiest tools for converting data from spreadsheet formats into line and polygon data layers. At this stage, it is probably most useful for those biologists who wish to simply make maps for presentations and publications, but as more tools are developed, it is likely to become increasingly useful for biologists.

SECTION TWO:

PRACTICAL EXERCISES

--- Chapter Eleven ---

How To Use The GIS Software User Interfaces

For the practical exercises in this book, you can choose between two different GIS software packages. The first is ArcGIS 10.3. This is the leading commercial GIS software package, and it requires a specific licence which may not be available to all those interested in learning about using GIS in biological research. Therefore, instructions for each exercise are also provided for QGIS 2.8.3. QGIS, or Quantum GIS, is a freely available, open source GIS software package. Information on where these two software packages can be obtained, and how to download and install them can be found on *www.GISinEcology.com/ GFB.htm#1*. You should select your preferred software package and install it on your computer. Details of how to use the user interfaces of these two packages are provided below.

ArcGIS 10.3 User Interface:

All of the instructions provided in this section of the book assume that you are working with the main ArcGIS 10.3 user interface, which is a module known as ArcMap. As a result, it is important to outline how the ArcMap window is laid out. The ArcMap user interface is divided into a number of sections and windows (figure 8). These sections include: a MAIN MENU BAR area that allows you to access some basic and commonly used tools through a variety of standard drop down menus; the OPTIONAL TOOLBARS AREA, where you can display toolbars for specific toolboxes to allow you to access them quickly and easily; the TABLE OF CONTENTS window, which displays a list of all the data layers in your GIS project; the TOOLBOX window, which allows you to access the various tools available in the software; an ADDITIONAL OPTIONAL TOOLBARS AREA; an X-Y COORDINATE DISPLAY area, which provides the X and Y coordinates for the position of the cursor in the MAP window; and the MAP window which displays all the active data layers within an active data frame.

When you open the ArcMap module for the first time, you may find that some of the windows (such as the TOOLBOX window) are not be visible at all. This is because it can be customised depending on the user preferences. However, for the purposes of this book, it is useful to standardise this layout in terms of what is visible at all times and where the different windows are placed. To do this, first check if the TABLE OF CONTENTS window is visible (see figure 8). If it is not, go to the WINDOWS menu and select TABLE OF CONTENTS. If it is visible (or once you have made it visible), click on its title bar and hold the mouse button down. Now, drag the TABLE OF CONTENTS window towards the middle of the MAP window. Four blue arrows will now be visible there. Drop the window onto the left hand one of these arrows. This will tie the TABLE OF CONTENTS window to the left hand side of the MAP window. Next, click on the GEOPROCESSING menu and select ARCTOOLBOX. The TOOLBOX window will now appear at the left hand side of the TABLE OF CONTENTS window. Move it to the right of this window, by clicking on its title bar and holding the mouse button down, before dragging it to the centre of the MAP window and dropping it onto the left hand blue arrow that will appear there. The ArcMap 10.3 user interface will then have a layout like that shown in figure 8. It is advisable that you set up the ArcMap 10.3 user interface to look like this before working through the practical exercises provided in this book.

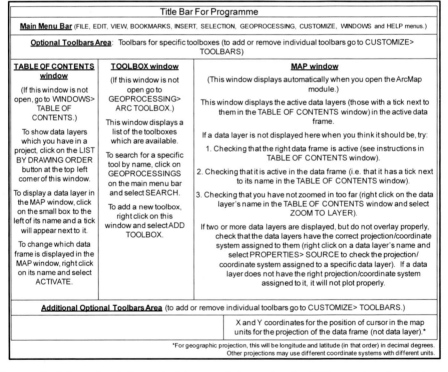

Figure 8. *Schematic of the typical layout of the main ArcGIS user interface (known as ArcMap). Within this book, the individual sections of the window will be referred to using the names which are <u>underlined</u> and in **bold**.*

NOTE: It is useful at this time to change two other default settings of the ArcMap module. Click on the GEOPROCESSING menu on the main menu bar and select GEOPROCESSING OPTIONS. Make sure that there is a tick next to OVERWRITE THE OUTPUTS OF GEOPROCESSING OPERATIONS. This will allow you to over-write any files you create if you find you have made a mistake and need to make them again (this is very common in GIS). You also need to make sure that there is no tick next to ENABLE under BACKGROUND PROCESSING in the GEOPROCESSING OPTIONS window. This will mean you can see what the software is doing at any particular moment, and will reduce the chances that you will miss operations which are being conducted in the background.

QGIS 2.8.3 User Interface:

When you first open QGIS, you will find that the main user interface is divided into a number of sections and windows. However, before you start working through the practical exercises in this book, you will need to standardise its appearance (see *www.GISinEcology.com/GFB.htm#Video11* for more information on how to do this). To do this, first click on the VIEW menu and select PANELS. Make sure that only the panels which are set to display are LAYERS and TOOLBOX. If any others are set to display, uncheck the box next to their names. Make sure that the panel titled LAYERS is on the left hand side and the one titled PROCESSING TOOLBOX is on the right. At the bottom of the PROCESSING TOOLBOX window, you have the option of selecting an ADVANCED INTERFACE or a SIMPLIFIED INTERFACE. Select the ADVANCED INTERFACE option. Next, you may find that you have one or more toolbars running vertically down the left hand side of the QGIS window. If this is the case, click on the stippled bar at the top and drag them to the top of the window, where the other toolbars are displayed. Finally, click on the VIEW menu again and select TOOLBARS. Make sure all the toolbars are set to display with the exception of ADVANCED DIGITIZING and DATABASE. The QGIS window should now look like figure 9 (**NOTE:** The exact appearance may vary slightly depending on where you have each individual toolbar displayed).

The QGIS user interface has a number of key areas (see *www.GISinEcology.com/GFB.htm#Video11* for more information). Along the very top of the window is a MAIN MENU BAR area, which allows you to access tools for creating and saving GIS projects (under the PROJECT menu), for changing what is displayed in the QGIS interface (under the VIEW menu), as well as the various editing and geoprocessing tools (under the remaining menus). Below the MAIN MENU BAR is an OPTIONAL TOOLBARS

AREA, where you can display toolbars for specific tool sets and plugins to allow you to access them quickly and easily. On the left hand side, there is a section called LAYERS which displays a list of all the data layers in your GIS project and it will be referred to as the TABLE OF CONTENTS window throughout this book. To the right of the TABLE OF CONTENTS window is the MAP window. This is where all the active data layers in your GIS project are displayed. On the right hand side is the PROCESSING TOOLBOX window. This window allows you to access a variety of tools, and will be referred to as the TOOLBOX window. At the bottom right hand corner is an X-Y COORDINATE DISPLAY area, which provides the X and Y coordinates for the position of the cursor in the MAP window, as well as tools for setting the scale of your MAP window, for deciding how your map is drawn and for setting the coordinate reference system (CRS) – including the CRS Status button. However, in this book, it will be referred to as the projection/coordinate system.

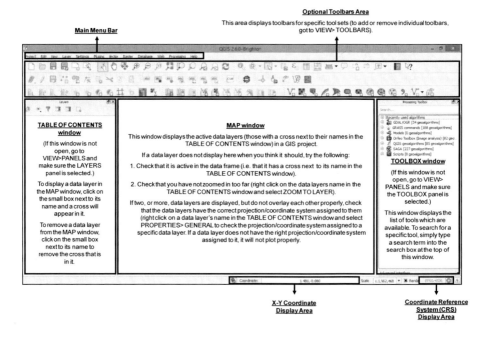

Figure 9. *The standard layout of the QGIS 2.8.3 user interface which will be assumed for the practical exercises in this book. Within this book, the individual sections of the window will be referred to using the names which are* <u>underlined</u> *and in* **bold**.

Chapter Twelve

Exercise One: How To Make Your First Map

One of the most common things for biologists to do in GIS is to make a map for inclusion in a report, presentation or publication. Such maps can be used to show things like where in the world a specific study was conducted, the distribution of sampling locations, the range of a species or the locations where a specific disease has been recorded. A good map will make it much easier for others to understand your work and for you to get your message across. However, creating a good map is not always as straight forward as it might, at first, seem. In particular, there is often a tendency to have either too much or too little information on your maps, to not show the right part of the world, or to miss out an important component which will help others understand what it is meant to show.

A good map should have the following characteristics:

- It should be easy to differentiate between symbols which represent different things (e.g. symbols representing the locations where different species have been recorded should not look too similar).

- The symbols should not be so large that they cover up other important features, or so small that you cannot tell what they are.

- You should be able to tell which part of the world it is representing. This can be done using a small insert map showing where the main map is situated in relation to a whole country, continent or the entire world, or by labelling the map with coordinates around its edge.

- You should be able to determine which direction is North. This can be done using a north arrow, or by labelling the map with coordinates around its edge.

- You should be able to easily work out how far apart any features are on it. This is usually done by adding a scale bar.

- It should show all the features you need to get your message across, but no more than that. If you have too few features, anyone viewing the map will not be able to understand what it represents, while if you have too many, it will become difficult to interpret.

- The colours used for different features on your map should compliment each other rather than clash, and colours should be used in a logical way. For example, avoid using green to represent areas of sea and blue for areas of land as this is likely to cause confusion. In addition, if your final map is going to be printed in greyscale (rather than in colour), make sure that you can still differentiate between all the colours once it has been converted into this colour scheme. If you cannot find an appropriate colour scheme that will work in greyscale, you can try using different fill patterns instead.

In this exercise, you will create your first map and you will learn how to add various components to it so that it is easy to understand and interpret. You will also learn how to change the way that individual features are displayed and the symbols that are used to represent them. The final map which you will generate would be suitable for inclusion in an scientific report or publication.

The map you will create is based on data collected along transects which have been established around the Manu Learning Centre (MLC), a field station in the Manu Biosphere Reserve in the Peruvian Amazon run by an organisation called CREES (see *www.crees-manu.org* for more information). These transects are used to collect information on the distribution of a wide variety of organisms in a protected area of rainforest which has been established beside the MLC. The map you will produce will show the location of the MLC, along with the limits of its associated protected area, a portion of the survey transects which are regularly surveyed, and the locations along these transects where three species of macaws (a group of colourful New World parrots) have been recorded. The map will be used to show that while all three species use the local rainforest, they do not occur at the same locations, suggesting that there is some sort of spatial segregation between them. Such a map would be the first step in investigating whether the suggested spatial segregation between these species actually exists.

The instructions for creating this map in ArcGIS 10.3 can be found on page 74, while those for creating the map in QGIS 2.8.3 can be found on page 90. However, before you start this exercise, you will first need to create a new folder on your C: drive called GIS_FOR_BIOLOGISTS. To do this on a computer with a Windows operating system, open Windows Explorer and navigate to your C:\ drive (this may be called Windows C:). To create a new folder on this drive, right click on the window displaying the contents of your C:\ drive and select NEW> FOLDER. Now call this folder GIS_FOR_ BIOLOGISTS by typing this into the folder name to replace what it is currently called (which will most likely be NEW FOLDER). This folder, which has the address C:\GIS_FOR_BIOLOGISTS, will be used to store all files and data for the exercises in this book.

Next, you need to download the source files for six existing data layers which will be used in this exercise (and, indeed, source files for all the other exercises in this book) from *www.gisinecology.com/GFB.htm#2*. Once you have downloaded the compressed folder containing the files, make sure that you then copy all the files it contains into the folder C:\GIS_FOR_BIOLOGISTS, which you have just created.

The data sets which you will use in this exercise are:

1. **MLC_Protected_Area.shp:** This is a polygon data layer with a single polygon in it which represents the area of protected rainforest associated with the Manu Learning Centre (MLC). It is in the UTM_19S projection and uses the WGS 1984 datum for its coordinate system.

2. **MLC_Transects.shp:** This is a line data layer where the lines represent the transects regularly surveyed for a wide variety of organisms in the Manu Learning Centre (MLC) protected area. It is in the UTM_19S projection and uses the WGS 1984 datum for its coordinate system.

3. **MLC_Location.shp:** This is a point data layer where the single point in it represents the location of the Manu Learning Centre (MLC) itself. It is in the UTM_19S projection and uses the WGS 1984 datum for its coordinate system.

4. **Red_And_Green_Macaw.shp, Blue_Headed_Macaw.shp and Blue_ And_Yellow_Macaw.shp:** These three data layers are point data layers where each point in it represents a location where each species has been recorded during the surveys. They are in the UTM_19S projection, which uses the WGS 1984 datum for its coordinate system.

Instructions For ArcGIS 10.3 Users:

Once you have the required files downloaded into the correct folder on your computer, and you understand what is contained within each file, you can move on to creating your map. The starting point for this is a blank GIS project. To create a blank GIS project, first, start the ArcGIS software by opening the ArcMap module. When it opens, you will be presented with a window which has the heading ARCMAP – GETTING STARTED. In this window you can either select an existing GIS project to work on, or create a new one. To create a new blank GIS project, click on NEW MAPS in the directory tree on the left hand side and then select BLANK MAP in the right hand section of the window. Now, click OK at the bottom of this window. This will open a new, blank GIS project. (**NOTE:** If this window does not appear, you can start a new project by clicking on FILE from the main menu bar area, and selecting NEW. When the NEW DOCUMENT window opens, select NEW MAPS and then BLANK MAP in the order outlined above.) Once you have opened your new GIS project, the first thing you need to do is save it under a new, and meaningful, name. To do this, click on FILE on the main menu bar, and select SAVE AS. Save it as EXERCISE_ONE in the folder C:\GIS_FOR_BIOLOGISTS.

STEP 1: SET THE PROJECTION AND COORDINATE SYSTEM OF YOUR DATA FRAME:

Whenever you start a new GIS project, the first thing you should do, even before you add any data layers, is select an appropriate projection/coordinate system and then set the data frame (i.e. the map you are going to add your data layers to) to use this projection/coordinate system. This ensures that you do not accidently end up using one which is inappropriate. In addition, most problems that beginners have with GIS are caused by a failure to use an appropriate projection/coordinate system, or by a conflict between the projection/coordinate system used for data layers and the one being used for the data frame itself. For this exercise you will use a pre-existing projection/coordinate system called Universal Transverse Mercator (UTM) Zone 19S. This is a transverse mercator projection which is specifically designed for the region of the world where the Manu Learning Centre (MLC) is situated.

To set your data frame to use the UTM Zone 19S projection/coordinate system, work through the flow diagram which starts on the next page.

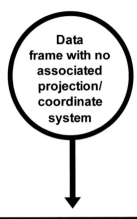

Data frame with no associated projection/ coordinate system

When you open a new GIS project, there will be an empty data frame (called, rather confusingly, LAYERS). It will not have a projection or coordinate system associated with it.

1. Open data frame properties

To open the properties of a data frame, right click on its name (in this case LAYERS) in the TABLE OF CONTENTS window, and select the PROPERTIES option. If this is not visible, go to the WINDOWS menu and click on the TABLE OF CONTENTS option.

2. Go to the coordinate system window

Select the COORDINATE SYSTEM tab in the DATA FRAME PROPERTIES window. The lower section of this window will tell you which projection and coordinate system are set for the data frame you are currently using. This should currently say 'No coordinate system'.

3. Select the required pre-existing projection/ coordinate system

To select a pre-existing projection/coordinate system, select it from the upper section of the COORDINATE SYSTEM tab. For this exercise, you wish to select WGS_1984_UTM_ZONE_19S (this is a suitable projection/coordinate system for the area of Peru which will be mapped in this exercise). To select it, click on PROJECTED COORDINATE SYSTEMS> UTM> WGS 1984> SOUTHERN HEMISPHERE> WGS 1984 UTM ZONE 19S (you may need to scroll down to find this option). Once you have selected it, its name will appear in the CURRENT COORDINATE SYSTEM section of the COORDINATE SYSTEM tab along with other information about this specifc projection/coordinate system. Finally, click OK to close the DATA FRAME PROPERTIES window.

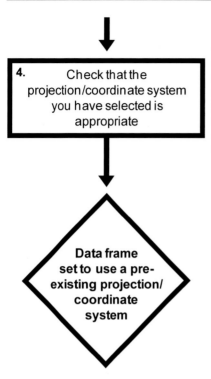

Once you have selected your projection/ coordinate system, you need to check that it is appropriate. This involves examining how data layers look in it. For this exercise, this will be done in the next step by adding a series of data layers, and checking that the features in them plot in the expected places and that any polygons have the expected shapes.

To check that you have done this step properly, right click on the name of your data frame (LAYERS) in the TABLE OF CONTENTS window and select PROPERTIES. Click on the COORDINATE SYSTEM tab of the DATA FRAME PROPERTIES window and make sure that the contents of the CURRENT COORDINATE SYSTEM section of the window has the following text at the top of it:

WGS_1984_UTM_ZONE_19S
WKID: 32719 Authority EPSG

Projection: Transverse_Mercator
False_Easting: 500000.0
False_Northing: 10000000.0
Central_Meridian: -69.0
Scale_Factor: 0.9996
Latitude_Of_Origin: 0.0
Linear Unit: Meter (1.0)

If it does not, you will need to repeat this step until you have assigned the correct projection/coordinate system to your data frame. Once you have successfully completed this step, click on the FILE menu on the main menu bar and select SAVE to save the changes you have made to your GIS project.

STEP 2: ADD THE EXISTING DATA LAYERS YOU WISH TO DISPLAY ON YOUR MAP TO YOUR GIS PROJECT:

Once you have set the projection/coordinate system of your data frame, you are ready to add some data layers to it. For this exercise, you will use a number of existing data layers which provide information about a variety of different features for the area around the Manu Learning Centre (MLC). These will provide information about the location of the area of protected rainforest associated with the MLC (contained in the polygon data layer MLC_PROTECTED_AREA), about a set of transect lines which are regularly surveyed in this protected area (contained in the line data layer MLC_TRANSECTS), the location of the MLC itself (contained in the point data layer MLC_LOCATION) and the locations where three different species of macaws have been recorded during the transect surveys (contained in the point data layers RED_AND_GREEN_MACAW, BLUE_HEADED_MACAW and BLUE_AND_YELLOW_MACAW).

To add these data layers to your GIS project, work through the following flow diagram:

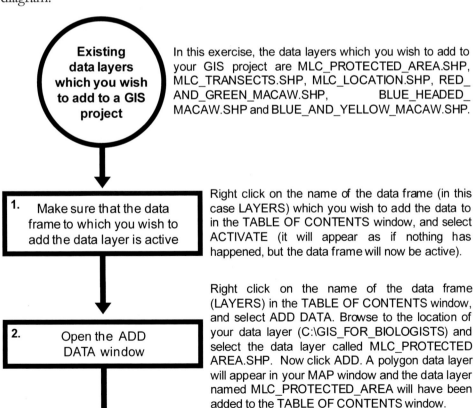

Existing data layers which you wish to add to a GIS project

In this exercise, the data layers which you wish to add to your GIS project are MLC_PROTECTED_AREA.SHP, MLC_TRANSECTS.SHP, MLC_LOCATION.SHP, RED_AND_GREEN_MACAW.SHP, BLUE_HEADED_MACAW.SHP and BLUE_AND_YELLOW_MACAW.SHP.

1. Make sure that the data frame to which you wish to add the data layer is active

Right click on the name of the data frame (in this case LAYERS) which you wish to add the data to in the TABLE OF CONTENTS window, and select ACTIVATE (it will appear as if nothing has happened, but the data frame will now be active).

2. Open the ADD DATA window

Right click on the name of the data frame (LAYERS) in the TABLE OF CONTENTS window, and select ADD DATA. Browse to the location of your data layer (C:\GIS_FOR_BIOLOGISTS) and select the data layer called MLC_PROTECTED AREA.SHP. Now click ADD. A polygon data layer will appear in your MAP window and the data layer named MLC_PROTECTED_AREA will have been added to the TABLE OF CONTENTS window.

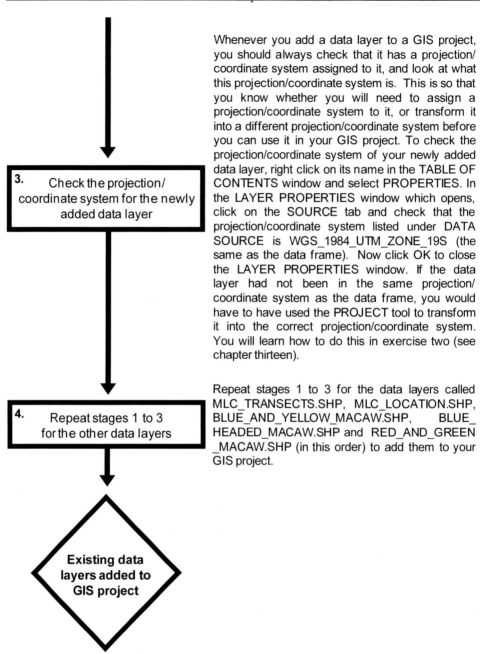

Whenever you add a data layer to a GIS project, you should always check that it has a projection/coordinate system assigned to it, and look at what this projection/coordinate system is. This is so that you know whether you will need to assign a projection/coordinate system to it, or transform it into a different projection/coordinate system before you can use it in your GIS project. To check the projection/coordinate system of your newly added data layer, right click on its name in the TABLE OF CONTENTS window and select PROPERTIES. In the LAYER PROPERTIES window which opens, click on the SOURCE tab and check that the projection/coordinate system listed under DATA SOURCE is WGS_1984_UTM_ZONE_19S (the same as the data frame). Now click OK to close the LAYER PROPERTIES window. If the data layer had not been in the same projection/coordinate system as the data frame, you would have to have used the PROJECT tool to transform it into the correct projection/coordinate system. You will learn how to do this in exercise two (see chapter thirteen).

Repeat stages 1 to 3 for the data layers called MLC_TRANSECTS.SHP, MLC_LOCATION.SHP, BLUE_AND_YELLOW_MACAW.SHP, BLUE_HEADED_MACAW.SHP and RED_AND_GREEN_MACAW.SHP (in this order) to add them to your GIS project.

At the end of this step, your TABLE OF CONTENTS window should look like the image at the top of the next page.

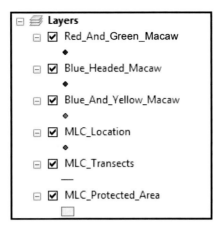

While the contents of your MAP window should look like this:

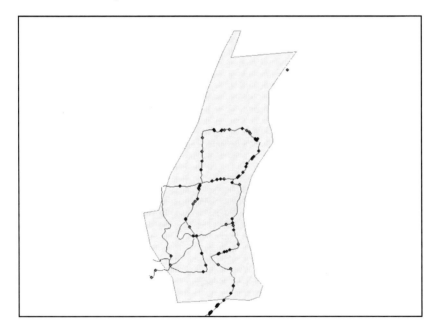

If it does not look like this, try right-clicking on the name of the MLC_PROTECTED_AREA data layer in the TABLE OF CONTENTS window and selecting ZOOM TO LAYER. If it still does not look right, remove all the data layers from your GIS project by right-clicking on their names in the TABLE OF CONTENTS window and selecting REMOVE. Next, go back to step one and ensure that you have set the projection/coordinate system of your data frame correctly, and then repeat step two. Once you have successfully completed this step, click on the FILE menu on the main menu bar and select SAVE to save the changes you have made to your GIS project.

STEP 3: SET HOW THE INFORMATION IN THE DATA LAYERS IN YOUR GIS PROJECT WILL BE DISPLAYED ON YOUR MAP:

Once you have added data layers to your GIS project, you will almost certainly need to change the way they display the information which they contain. This can involve changing the size and colour of the symbols used, changing the shading of polygons, changing the thickness of lines, deciding whether to use a single symbol for all the features in a data layer or whether you will use different symbols for different classes of features, and deciding if you need to add labels to help anyone looking at your map identify what specific features represent.

To do this for the six data layers you added to your GIS project in step two, work through the following flow diagram:

Data layers in a GIS project which you wish to display in a different way

In this exercise, the data layers which you wish to display in a different way are MLC_PROTECTED_AREA, MLC_TRANSECTS, MLC_LOCATION, BLUE_AND_YELLOW_MACAW, BLUE_HEADED_MACAW and RED_AND_GREEN_MACAW.

1. Select how the MLC_PROTECTED_AREA data layer will be displayed

Right click on the name of the MLC_PROTECTED_AREA data layer in the TABLE OF CONTENTS window, and select PROPERTIES. In the LAYER PROPERTIES window, click on the SYMBOLOGY tab. On the left hand side under SHOW, select FEATURES> SINGLE SYMBOL and then click on the coloured box under SYMBOL. In the SYMBOL SELECTOR window which will open, click on the box next to FILL COLOR and select GRAY 10% (the second colour down in the left hand column of colours). Now click on the box next to OUTLINE WIDTH and enter 2. Click OK to close the SYMBOL SELECTOR window and then OK to close the LAYER PROPERTIES window. You will see that the way the MLC_PROTECTED_AREA data layer is displayed in the MAP window has now changed to the new settings you have just selected.

2. Select how the MLC_TRANSECTS data layer will be displayed

Right click on the name of the MLC_TRANSECTS data layer in the TABLE OF CONTENTS window, and select PROPERTIES. In the LAYER PROPERTIES window, click on the SYMBOLOGY tab. On the left hand side under SHOW, select FEATURES> SINGLE SYMBOL and then click on the box with the coloured line on it under SYMBOL. In the SYMBOL SELECTOR window which will open, click on the box next to COLOR and select POINSETTIA RED (the fourth colour down in the column of red shades). Now click on the box next to WIDTH and enter 2. Click OK to close the SYMBOL SELECTOR window and then OK to close the LAYER PROPERTIES window. You will see that the way the MLC_TRANSECTS data layer is displayed in the MAP window has now changed to the new settings you have just selected.

3. Select how the MLC_LOCATION data layer will be displayed

Right click on the name of the MLC_LOCATION data layer in the TABLE OF CONTENTS window, and select PROPERTIES. In the LAYER PROPERTIES window, click on the SYMBOLOGY tab. On the left hand side under SHOW, select FEATURES> SINGLE SYMBOL and then click on the box with the coloured circle on it under SYMBOL. In the SYMBOL SELECTOR window which will open, click on SQUARE 1 to select this symbol and then for SIZE, enter the value 15.00. Click OK to close the SYMBOL SELECTOR window and then OK to close the LAYER PROPERTIES window. You will see that the way the MLC_LOCATION data layer is displayed in the MAP window has now changed to the new settings you have just selected.

4. Add a label to the symbol for the MLC_ LOCATION data layer

You now want to add a label to the symbol you have selected so that anyone looking at the map will know this is the location of the Manu Learning Centre. To do this, right click on the name MLC_LOCATION in the TABLE OF CONTENTS window and select PROPERTIES again. In the LAYER PROPERTIES window which will open, click on the LABELS tab. Next, select LOCATION in the LABEL FIELD section and then in the TEXT SYMBOL section, select TIMES NEW ROMAN for the font and 12 for the font size. Now click on PLACEMENT PROPERTIES and click on the CHANGE LOCATION button. In the INITIAL POINT PLACEMENT window which opens, scroll right down to the bottom and select PREFER BOTTOM RIGHT, ALL ALLOWED and then click OK to close this window. Click OK to close the PLACEMENT PROPERTIES window and then OK to close the LAYER PROPERTIES window. Now you have selected the settings for your label, you need to add it to your map. To do this, right click on MLC_LOCATION in the TABLE OF CONTENTS window and select LABEL FEATURES. The label will now be added to the MAP window beside the symbol for the MLC_ LOCATION. This label is based on the contents of the LOCATION field in the attribute table of this data layer.

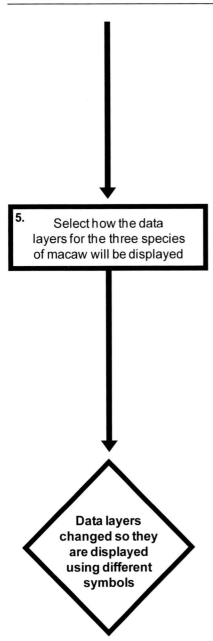

5. Select how the data layers for the three species of macaw will be displayed

Data layers changed so they are displayed using different symbols

Right click on the name of the RED_AND_GREEN_MACAW data layer in the TABLE OF CONTENTS window, and select PROPERTIES. In the LAYER PROPERTIES window, click on the SYMBOLOGY tab. On the left hand side under show, select SINGLE SYMBOL and then click on the box with the coloured circle on it under SYMBOL. In the SYMBOL SELECTOR window which will open, click on CIRCLE 2 to select this symbol and then for SIZE enter the value 12. Next, click on the box next to COLOR and select MEDIUM APPLE (it is the third colour down in the right hand column of green shades). Click OK to close the SYMBOL SELECTOR window, and then OK to close the LAYER PROPERTIES window. You will see that the way the RED_AND_GREEN_MACAW data layer is displayed in the MAP window has now changed to the new settings you have just selected.

Repeat this for the BLUE_HEADED_MACAW data layer, but for COLOR, select LAPIZ LAZULI (the fourth colour down in the right hand column of blue shades). Finally, repeat this for the BLUE_ AND_YELLOW_MACAW data layer, but for COLOUR, select SOLAR YELLOW (the third colour down in the column of yellows).

Once you have completed this instruction set, your TABLE OF CONTENTS window should look like the image at the top of the next page.

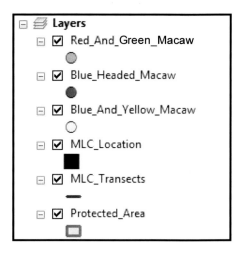

And the contents of your MAP window should look like this:

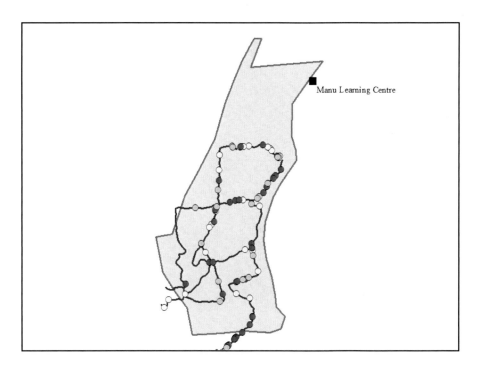

If it does not look like this, then work through this step again and ensure that you have set all the data layers to be displayed in the required ways. Once you have successfully completed this step, click on the FILE menu on the main menu bar and select SAVE to save the changes you have made to your GIS project.

STEP 4: CREATE AND EXPORT YOUR MAP:

Now that you have all the required data layers added to your GIS project, you can get an idea of how your final map will look. It is unlikely that you will get your map looking perfect on the first attempt, and you will almost always have to change something. The instructions for this step are relatively simple and so are provided as text without an accompanying flow diagram. Images are provided along the way to allow you to check that you have completed each section correctly.

Open the LAYOUT window by going to the VIEW menu on the main menu bar and selecting LAYOUT VIEW. (If you want to get back to the MAP window at any point, simply go to the VIEW menu on the main menu bar and select DATA VIEW.) In the LAYOUT window (which will replace the MAP window), you will see what your map will look like for your specific data frame. When you initially look at it, you will undoubtedly not be too impressed. However, with a few simple steps, you can make it look much better.

The first thing that you will need to do is make sure that the units that are being used for the LAYOUT VIEW are centimetres and not inches. To do this, click on CUSTOMIZE on the main menu bar and select ARCMAP OPTIONS. In the OPTIONS window, click on the LAYOUT VIEW tab, and then in the RULERS section, select CENTIMETRES from the drop down menu next to UNITS. Finally, click OK to close the OPTIONS window.

The next thing you need to do is to change the extent of the data frame so that your map only shows the area you want it to. To do this, right-click on the name of your data frame (LAYERS) in the TABLE OF CONTENTS window and select PROPERTIES. In the DATA FRAME PROPERTIES window, select the DATA FRAME tab. Here, you will find the EXTENT options. To limit the extent of your map, select FIXED EXTENT from the drop down menu. You can then enter the required coordinates for the limits of the extent by filling in the options that appear. In this exercise, you want to focus in on the areas covered by the MLC protected area. This can be done by setting the TOP limit to 8587370, the LEFT limit to 238190, the RIGHT limit to 241900, and the BOTTOM limit to 8581400. Once you have entered these values, click the APPLY button at the bottom of the DATA FRAME PROPERTIES window and you will see the extent of your map will change in the LAYOUT window.

If you did not already have these coordinates, you could get an idea of what coordinates would be appropriate to define the extent for a specific map before you open the DATA FRAME PROPERTIES window by going to the MAP window and using the ZOOM IN or ZOOM OUT tools until you can see all the data you wish to show on your map. Next, move the cursor to the top left of the area you wish your map to show. You can then read off the coordinates which you would need to use to set the left (the first coordinate) and top (the second coordinate) limits for the extent of your map by looking at the coordinate display area at the bottom right-hand corner of the ArcMap user interface (see figure 8 on page 68). Repeat this for the bottom right-hand corner of the area you wish your map to show and read off the right (the first coordinate) and bottom (the second coordinate) limits for the extent of your map. Once your have these values, you can set the extent of your map as outlined above. If you want to give this a go, you would first need to change the extent setting from FIXED EXTENT to AUTOMATIC to allow you to zoom in and out in the MAP window. However, remember to set the extent back to a fixed extent using the above coordinates before you continue with this exercise. If you do not, the contents of the LAYOUT window and the final map will not match the images provided.

Next, you want to set the size and the position of the window. Select the SIZE AND POSITION tab of the DATA FRAME PROPERTIES window. For POSITION, enter 3cm for X and 3cm for Y and click APPLY, you will see the position of your map change. Finally, for size enter 15cm for WIDTH and then click on the HEIGHT box. This will automatically update the HEIGHT value to ~24cm to maintain the aspect ratio of the data frame (which is set by the extent coordinates you entered above). You can now click OK to close the DATA FRAME PROPERTIES window and have a closer look at your map in the LAYOUT window. It should now look like this:

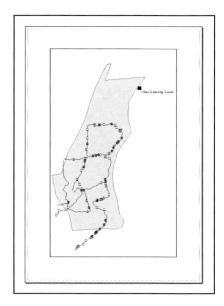

If it does not look like this, go back and check that you set the extent, size and position to the correct values given in the instructions before carrying on.

The next thing to do is to add a latitude and longitude grid around the edges of your map so that people will know what part of the world it represents. As with the extent, size and position, this is also done in the DATA FRAME PROPERTIES window. Open this window again by right-clicking on the name of your data frame (LAYERS) in the TABLE OF CONTENTS window and selecting PROPERTIES. This time click on the GRIDS tab. Now click on the NEW GRID button. This will open the GRIDS AND GRATICULES wizard, and you can work your way through the options to get your latitude and longitude grid looking exactly the way that you want. For the purposes of this exercise, you will mostly use the default settings for your latitude and longitude grid, but there are many options which you can use to customise how your latitude and longitude grid will look. Once you have a bit more experience you can play around with all the options until you find ones you like.

Firstly, in the GRIDS AND GRATICULES wizard, you will select GRATICULES: DIVIDES MAP BY MERIDIANS AND PARALLELS and then click the NEXT button. This will take you to the CREATE A GRATICULE window. Select LABELS ONLY and set the INTERVALS for both parallels and meridians to 0 DEG, 0 MIN and 30 SEC. Then click the NEXT button. This will take you to the AXES AND LABELS window. Click on the TEXT STYLE button and in the SYMBOL SELECTOR window which opens, set the SIZE to 12, and select TIMES NEW ROMAN for the font, and then click OK. Click the NEXT button in the AXES AND LABELS window. Finally, click the FINISH button to close the CREATE A GRATICULE window, and then click OK to close the DATA FRAME PROPERTIES window.

Next, you will add a scale bar. To do this, click on the INSERT menu on the main menu bar and select SCALE BAR. This will open the SCALE BAR SELECTOR window. You will find there are many different ways you can set your scale bar, but for this exercise, you will use a relatively simple one. Click on the top scale bar design option on the left hand side of the SCALE BAR SELECTOR window, then click on the PROPERTIES button to open the SCALE BAR window. Here you can change the settings to get exactly the right design of scale bar you wish. For this exercise, you will simply use all the default options, with the exception of the DIVISION UNITS option. For this option, you will change the units from MILES to KILOMETRES. Once you have done this, click OK to close the SCALE BAR window, and then OK to close the SCALE BAR SELECTOR window. You will see a scale bar has now been added to your map in the LAYOUT window. The first thing to do is to change its size, if required. This is done by double-clicking on the scale bar to open the SCALE LINE PROPERTIES window and selecting the SIZE AND POSITION tab. For this exercise, enter a value of 6.25 cm for the WIDTH and then click OK. This will set the scale bar to a reasonable length for your map. Finally, you need to position the scale bar where you want it to be. For this exercise, you will want to move the scale bar down to the lower right hand corner of the map. This is done by clicking on it and then dragging it while holding the left hand mouse button down. The contents of your LAYOUT window should now look like this:

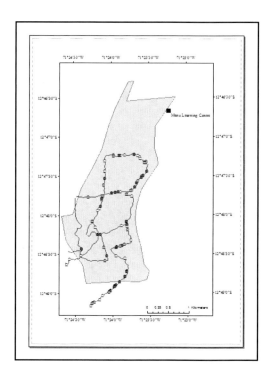

Once you have successfully inserted a scale bar, you will want to insert a North arrow to indicate which direction is north on your map. To do this, click on INSERT on the main menu bar and select NORTH ARROW. In the NORTH ARROW SELECTOR window which will open, select ESRI NORTH 3 and then click OK. This will add the selected North arrow to the map. The first thing to do is to change its size, if required. This is done by double-clicking on the North arrow to open the NORTH ARROW PROPERTIES window and selecting the SIZE AND POSITION tab. For this exercise, enter a value of 1 cm for the WIDTH and then click OK. This will set the North arrow to a reasonable size for your map. Finally, you need to position the North arrow where you want it to be. For this exercise, you will want to move the North arrow up to the top right hand corner of the map. This is done by clicking on it and then dragging it while holding the left hand mouse button down. The contents of your LAYOUT window should now look like the image at the top of the next page.

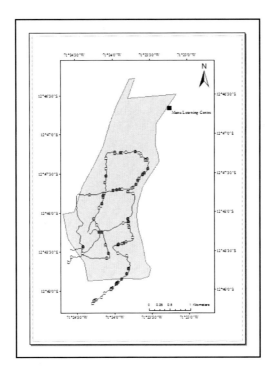

Since this is a relatively simple map, you can describe its contents in a figure legend which will tell the viewer what the different symbols and colours mean. For more complex maps, you might want to add a specific legend. This can be done through the INSERT menu.

All that is left now is for you to export your map using the EXPORT MAP tool. This is selected by clicking on the FILE menu on the main menu bar and selecting EXPORT MAP. This will open the EXPORT MAP window where you can select the format and resolution you wish to export the map in and the location where you wish to save it. The format and the resolution you select will depend on what you wish to use your map for. For example, you may choose a different format and resolution depending on whether you are creating a map to include in a presentation, or one for a written report. For this exercise, export your map as a .jpg with a resolution of 300 dpi, which would be suitable for most reports and manuscripts, call it MAP_ EXERCISE_ONE and save it in the folder C:\GIS_FOR_BIOLOGISTS. Once you have successfully completed this step, click on the FILE menu on the main menu bar and select SAVE to save the changes you have made to your GIS project.

Instructions For QGIS 2.8.3 Users:

Once you have the required files downloaded into the correct folder on your computer, and you understand what is contained within each file, you can move on to creating your map. The starting point for this is a blank GIS project. To create a blank GIS project, first open QGIS. By default, it will open with a blank GIS project. Once it is open, click on the PROJECT menu and select SAVE AS. In the window which opens, save your GIS project as EXERCISE_ONE in the folder C:\GIS_FOR_BIOLOGISTS.

STEP 1: SET THE PROJECTION AND COORDINATE SYSTEM OF YOUR DATA FRAME:

Whenever you start a new GIS project, the first thing you should do, even before you add any data layers, is select an appropriate projection and/coordinate system and then set the data frame (i.e. the map you are going to add your data layers to) to use this projection/coordinate system. This ensures that you do not accidently end up using one which is inappropriate. In addition, most problems that beginners have with GIS are caused by a failure to use an appropriate projection/coordinate system or by a conflict between the projection/coordinate system used for data layers and the one being used for the data frame itself. For this exercise, you will use a pre-existing projection/coordinate system called Universal Transverse Mercator (UTM) Zone 19S. This is a transverse mercator projection which is specifically designed for the region of the world where the Manu Learning Centre (MLC) is situated.

To set your data frame to use the UTM Zone 19S projection/coordinate system, work through the flow diagram which starts on the next page.

NOTE: If, while working through the instructions for this exercise, you have problems finding the CRS STATUS button in QGIS, you will find more information about where it is located in figure 9 on page 70. Alternatively, you can find a video which will help you locate it at *www.GISinEcology.com/ GFB.htm#Video11.*

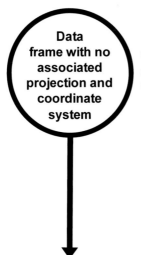

Data frame with no associated projection and coordinate system

When you open a new GIS project, there will be an empty data frame. It will not have a projection or coordinate system associated with it.

In the main QGIS window, click on the CRS STATUS button in the bottom right hand corner (it is square with a dark circular design on it). This will open the PROJECT PROPERTIES CRS window. Click on the box next to ENABLE 'ON THE FLY' CRS TRANSFORMATION so that a cross appears in it. This allows you to select a specific projection/coordinate system for your data frame. It also allows you to have data layers in different projection/coordinate systems in your GIS project and still have them plot in the correct places relative to each other. For biological GIS projects, you will almost always want to select this option.

1. Set your data frame to use a pre-existing projection/ coordinate system

Next, type the name of the projection/coordinate system that you want to use into the FILTER section (in this case, it will be WGS 84 / UTM ZONE 19S – **NOTE**: You will have to enter this name exactly as written, including the spaces, in order to be able to find it). In the COORDINATE REFERENCE SYSTEMS OF THE WORLD section, click on WGS 84 / UTM ZONE 19S under PROJECTED COORDINATE SYSTEMS> UNIVERSAL TRANSVERSE MERCATOR (UTM). This will add this to the SELECTED CRS section further down the window. Now click OK to close the PROJECT PROPERTIES CRS window.

2. Check that the projection/coordinate system you have selected is appropriate

Once you have selected a projection/coordinate system, you need to check that it is appropriate. This involves examining how data layers look in it. For this exercise, this will be done in the next step by adding a series of data layers and checking that the features in them plot in the expected places and that any polygons have the expected shapes.

Data frame set to use a pre-existing projection/ coordinate system

To check that you have done this step properly, click on PROJECT on the main menu bar and select PROJECT PROPERTIES. In the PROJECT PROPERTIES window which opens, click on the CRS tab on the left hand side and make sure that the contents of the SELECTED CRS section of the window has the following text in it:

WGS 84 / UTM ZONE 19S

Underneath this it should say:

+proj=utm +zone=19 +south +datum=WGS84 +units=m +no_defs

This is known as the Proj.4 string, which is used to define the characteristics of the projection/coordinate system.

If it does, click OK to close the PROJECT PROPERTIES window. If it does not, you will need to repeat this step until you have assigned the correct projection/coordinate system to your data frame. Once you have successfully completed this step, click on the PROJECT menu on the main menu bar and select SAVE to save the changes you have made to your GIS project.

STEP 2: ADD THE EXISTING DATA LAYERS YOU WISH TO DISPLAY ON YOUR MAP TO YOUR GIS PROJECT:

Once you have set the projection/coordinate system of your data frame, you are ready to add some data layers to it. For this exercise, you will use a number of existing data layers which provide information about a variety of different features for the area around the Manu Learning Centre (MLC). These will provide information about the location of the area of protected rainforest associated with the MLC (contained in the polygon data layer MLC_PROTECTED_AREA), about a set of transect lines which are regularly surveyed in this protected area (contained in the line data layer MLC_TRANSECTS), the location of the MLC itself (contained in the point data layer MLC_LOCATION) and the locations where three different species of macaws have been recorded during the transect surveys (contained in the point data layers RED_AND_GREEN_MACAW, BLUE_HEADED_ MACAW and BLUE_AND_YELLOW_MACAW).

To add these data layers to your GIS project, work through the flow diagram which starts on the next page.

Existing data layers which you wish to add to a GIS project

In this case, the data layers which you wish to add to your GIS project are MLC_PROTECTED_AREA.SHP, MLC_TRANSECTS.SHP, MLC_LOCATION.SHP, RED_AND_GREEN_MACAW.SHP, BLUE_HEADED_MACAW.SHP and BLUE_AND_YELLOW_MACAW.SHP.

1. Open the ADD VECTOR LAYER window

On the main menu bar, click on LAYER and select ADD LAYER> ADD VECTOR LAYER. In the ADD VECTOR LAYER window, browse to the location of your data layer (C:\GIS_FOR_BIOLOGISTS) and click on the section in the bottom right hand corner of the OPEN window (where it currently says ALL FILES (*).(*)) and select ESRI SHAPEFILES (*.shp, .SHP). Now, select the data layer called MLC_PROTECTED_AREA.SHP. Next, click OPEN in the browse window and then OPEN in the ADD VECTOR LAYER window.

2. Check the projection/coordinate system for the newly added data layer

Whenever you add a data layer to a GIS project, you should always check that it has a projection/coordinate system assigned to it, and look at what this projection/coordinate system is. This is so that you know whether you will need to assign a projection/coordinate system to it, or transform it into a different projection/coordinate system before you can use it in your GIS project. To check the projection/coordinate system of your newly added data layer, right click on its name in the TABLE OF CONTENTS window and select PROPERTIES. In the LAYER PROPERTIES window which opens, click on the GENERAL tab on the left hand side and check that there is a projection/coordinate system listed in the COORDINATE REFERENCE SYSTEM section. For the MLC_PROTECTED_AREA data layer this should say:

EPSG:32719 – WGS 84 / UTM ZONE 19S

This tells you that this data layer is in the UTM Zone 19S projection based on the WGS 1984 datum. Now click OK to close the LAYER PROPERTIES window.

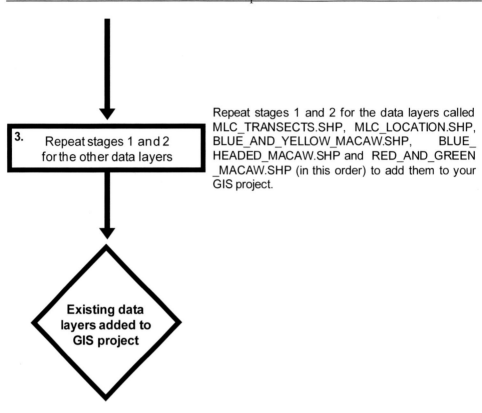

3. Repeat stages 1 and 2 for the other data layers

Repeat stages 1 and 2 for the data layers called MLC_TRANSECTS.SHP, MLC_LOCATION.SHP, BLUE_AND_YELLOW_MACAW.SHP, BLUE_ HEADED_MACAW.SHP and RED_AND_GREEN _MACAW.SHP (in this order) to add them to your GIS project.

Existing data layers added to GIS project

At the end of this step, your TABLE OF CONTENTS window should look like this (**NOTE:** If your Red_And_Green_Macaw data layer is not underlined, this is okay and can be safely ignored):

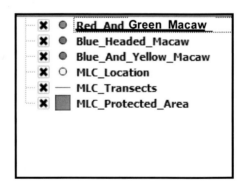

If your data layers are not in the same order as the above figure, you need to re-arrange them until they are. To do this, click on the name of a data layer, and while holding down the left hand mouse button, drag it upwards or downwards to the desired position.

The contents of your MAP window should look like this:

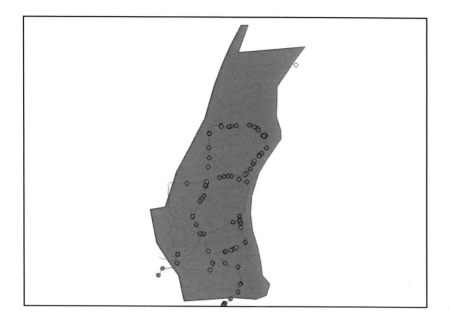

If it does not look like this, try right-clicking on the name of the MLC_PROTECTED_AREA data layer in the TABLE OF CONTENTS window and selecting ZOOM TO LAYER. If it still does not look right, remove all the data layers from your GIS project by right-clicking on their names in the TABLE OF CONTENTS window and selecting REMOVE. Next, go back to step one and ensure that you have set the projection/coordinate system of your data frame correctly, and then repeat step two. Once you have successfully completed this step, click on the PROJECT menu on the main menu bar and select SAVE to save the changes you have made to your GIS project.

STEP 3: SET HOW THE INFORMATION IN THE DATA LAYERS IN YOUR GIS PROJECT WILL BE DISPLAYED ON YOUR MAP:

Once you have added data layers to your GIS project, you will almost certainly need to change the way they display the information which they contain. This can involve changing the size and colour of the symbols used, changing the shading of polygons, changing the thickness of lines, deciding whether to use a single symbol for all the features in a data layer or whether you will use different symbols for different classes of features, and deciding if you need to add labels to help anyone looking at your map identify what specific features represent.

To do this for the six data layers you added to your GIS project in step two, work through the following flow diagram:

Data layers in a GIS project which you wish to display in a diferent way

In this case, the data layers which you wish to display in a different way are MLC_PROTECTED_AREA, MLC_TRANSECTS, MLC_LOCATION, BLUE_ AND_YELLOW_MACAW, BLUE_HEADED_MACAW and RED_AND_GREEN_MACAW.

1. Select how the MLC_PROTECTED_AREA data layer will be displayed

Right click on the name of the MLC_PROTECTED_AREA data layer in the TABLE OF CONTENTS window, and select PROPERTIES. In the LAYER PROPERTIES window, click on the STYLE tab on the left hand side. Next, in the top left hand corner of the STYLE tab, select SINGLE SYMBOL from the drop down menu, and then click on FILL> SIMPLE FILL in the section below it. Now on the right hand side of the window, click on the box next to COLORS FILL. In the SELECT FILL COLOR window which will open, select a grey colour by clicking on the appropriate shade on the colour selector on the left hand side and then click OK to close the SELECT COLOR window. Now enter 1.00 for BORDER WIDTH and then click OK to close the LAYER PROPERTIES window. You will see that the way the MLC_PROTECTED_AREA data layer is displayed in the MAP window has now changed to the new settings you have just selected.

2. Select how the MLC_TRANSECTS data layer will be displayed

Right click on the name of the MLC_TRANSECTS data layer in the TABLE OF CONTENTS window, and select PROPERTIES. In the LAYER PROPERTIES window, click on the STYLE tab on the left hand side. Next, select SINGLE SYMBOL from the drop down menu in the top left hand corner of the STYLE tab, and then click on LINE> SIMPLE LINE in the section below it. Now on the right hand side of the window, click on the box next to COLOR. In the SELECT LINE COLOR window which will open, select a red shade and click OK. Enter 1.00 for PEN WIDTH and then click OK to close the LAYER PROPERTIES window. You will see that the way the MLC_TRANSECTS data layer is displayed in the MAP window has now changed to the new settings you have just selected.

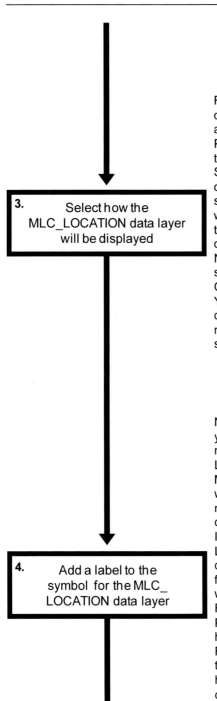

3. Select how the MLC_LOCATION data layer will be displayed

4. Add a label to the symbol for the MLC_LOCATION data layer

Right click on the name of the MLC_LOCATION data layer in the TABLE OF CONTENTS window, and select PROPERTIES. In the LAYER PROPERTIES window, click on the STYLE tab on the left hand side. In the top left hand corner on the STYLE tab, select SINGLE SYMBOL, and then click on MARKER> SIMPLE MARKER in the section below it. Now, on the right hand side of the window, click on the box next to COLORS FILL. In the SELECT FILL COLOR window which will open, select black as the colour and then click OK. Now enter 4.00 for SIZE and select the SQUARE symbol from the options available before clicking OK to close the LAYER PROPERTIES window. You will see that the way the MLC_LOCATION data layer is displayed in the MAP window has now changed to the new settings you have just selected.

Next, you now want to add a label to the symbol you have selected so that anyone looking at the map will know this is the location of the Manu Learning Centre. To do this, click on the name MLC_LOCATION in the TABLE OF CONTENTS window to select it. Next, click on LAYER on the main menu bar and select LABELING. This will open the LAYER LABELING SETTINGS window. In this window, check the box beside LABEL THIS LAYER WITH and then select LOCATION from the drop down menu next to it. Now click on TEXT from the menu on the lower left hand side of the window and then select TIMES NEW ROMAN for FONT and 11.0 for SIZE. Next, click on PLACEMENT from the menu on the lower left hand side of the window then select OFFSET FROM POINT. Select the lower right hand box in the QUADRANT section and enter 3.00 in the left hand box beside OFFSET X,Y. Finally, click OK to close the LAYER LABELING SETTINGS window. The label will now be added to the MAP window beside the symbol for the MLC_LOCATION. This label is based on the contents of the LOCATION field in the attribute table of this data layer.

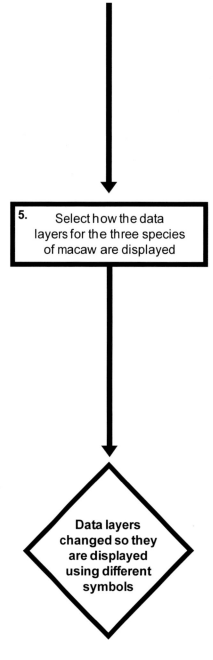

5. Select how the data layers for the three species of macaw are displayed

Data layers changed so they are displayed using different symbols

Right click on the name of the RED_AND_ GREEN_MACAW data layer in the TABLE OF CONTENTS window, and select PROPERTIES. In the LAYER PROPERTIES window, click on the STYLE tab on the left hand side. In the top left hand corner of the STYLE tab, select SINGLE SYMBOL and then click on MARKER> SIMPLE MARKER in the section below it. Now on the right hand side of the window, click on the box next to COLORS FILL. In the SELECT FILL COLOR window which will open, select a green shade and click OK. Now enter 3.00 for SIZE and select the CIRCLE symbol from the options available before clicking OK to close the LAYER PROPERTIES window. You will see that the way the RED_ AND_GREEN_MACAW data layer is displayed in the MAP window has now changed to the new settings you have just selected.

Repeat this for the BLUE_HEADED_MACAW data layer using the same symbol and size, but for COLOR select a blue shade. Finally, repeat this for the BLUE_AND_YELLOW_MACAW data layer, but for COLOR select a yellow shade.

Once you have completed this step, your TABLE OF CONTENTS window should look like the image at the top of the next page.

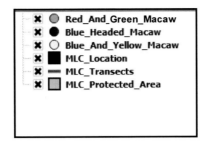

And the contents of your MAP window should look like this:

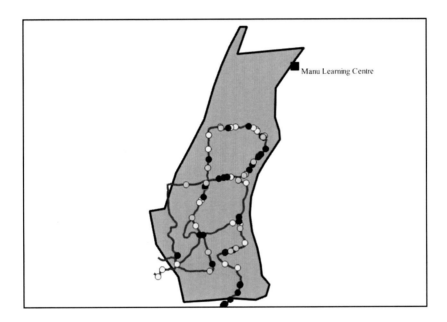

If it does not look like this, then work through this step again and ensure that you have set the data layers to be displayed in the required ways. Once you have successfully completed this step, click on the PROJECT menu on the main menu bar and select SAVE to save the changes you have made to your GIS project.

STEP 4: CREATE AND EXPORT YOUR MAP:

Now that you have all the required data layers added to your GIS project, you can get an idea of how your final map will look. It is unlikely that you will get your map looking perfect on the first attempt, and you will almost always have to change something. The instructions for this step are relatively simple and so are provided as text without an accompanying flow diagram. Images are provided along the way to allow you to check that you have completed each section correctly.

In QGIS, maps are created in the PRINT COMPOSER window. To open the PRINT COMPOSER window, click on PROJECT on the main menu bar and select NEW PRINT COMPOSER. A COMPOSER TITLE window will appear where you can enter a name for your map. Type in the name MLC MACAW MAP and then click OK. When the PRINT COMPOSER window first opens, it will be blank (this is okay) and it will be titled MLC MACAW MAP. The first thing you will do is set the paper size and orientation. To do this, click on the COMPOSITION tab on the right hand side of the PRINT COMPOSER window, then under PAPER AND QUALITY select A4 for PAPER SIZE, MM for UNITS, PORTRAIT for ORIENTATION and 300 DPI for EXPORT RESOLUTION (this value will define the resolution of your final map, and it is this value you will need to change if you want to alter your map's resolution). Leave all the other sections with their default settings. **NOTE:** If the COMPOSITION tab is not visible, click on VIEW on the main menu bar and select PANELS> COMPOSITION.

To add the contents of your MAP window to the PRINT COMPOSER window, click on LAYOUT on its main menu bar and select ADD MAP. Now move your cursor to the top left hand corner of the white box in the middle of the PRINT COMPOSER window. Hold the left hand button of your mouse down, move the cursor diagonally to the bottom right hand corner and then release it. The contents of your MAP window should now appear in the PRINT COMPOSER window. When you initially look at it, you will undoubtedly not be too impressed. However, with a few simple steps, you can make it look much better.

The first thing that you will need to do is set the size and the position of your map. To do this, click on the ITEM PROPERTIES tab of the right hand section of the PRINT COMPOSER window. Here, you can change various settings for your map. In this section, scroll down to the POSITION AND SIZE section and click on the black arrow to open the position and size options. For PAGE, enter 1, and then for POSITION, enter 40 for X and 50 for Y, and you will see the position of your map change. Now, enter 130 for WIDTH and 194 for HEIGHT. Next, you will need to set the extent so that your map only shows the area you want it to. In the ITEM PROPERTIES tab, scroll up until you find the section called EXTENTS, and enter 238190 for X MIN, 8581400 for Y MIN, 242200 for X MAX and 8587370 for Y MAX. **NOTE:** If you did not already have these coordinates, you could get an idea of what coordinates would be appropriate to define the extent for a specific map before you open the PRINT COMPOSER window by going to the MAP window and using the ZOOM IN or ZOOM OUT tools until you can see all the data you wish to show on your map. Next, move the cursor to the bottom left of the area you wish your map to show. You can then read off the coordinates which you would need to use to set the left, or X MIN, (the first coordinate) and bottom, or Y MIN, (the second coordinate) limits for the extent of your map by looking at the coordinate display area at the bottom of the QGIS user interface (see figure 9 on page 70). Repeat this for the top right hand corner of the area you wish your map to show and read off the right, or X MAX, (the first coordinate) and top, or Y MAX, (the second coordinate) limits for the extent of you map. Once you have these values, you can set the extent of your map as outlined above.

The contents of your PRINT COMPOSER window should now look like this (**NOTE:** You may have to scroll the window with the map in it up or down to see it properly):

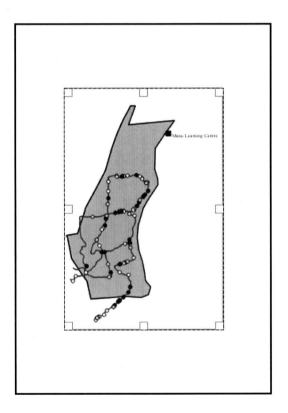

If it does not look like this, go back and check that you set the extent, size and position to the correct values given in the instructions before carrying on.

The next thing to do is to add a grid to your map so that people will know what part of the world it represents. As with the extent, size and position, this is also done in the ITEM PROPERTIES tab. In this tab, click on the black arrow beside GRIDS to reveal the GRID options, and then click on the +

button to create your grid. Now, scroll down and click on the button with CHANGE on it next to CRS. This will open the COORDINATE REFERENCE SYSTEM SELECTOR window. In this window, enter WGS 84 into the FILTER section (making sure that you leave a space between the letters WGS and the number 84), and then in the COORDINATE REFERENCE SYSTEMS OF THE WORLD section, under GEOGRAPHIC COORDINATE SYSTEMS, select the WGS 84 with the AUTHORITY ID of EPSG: 4326 (**NOTE:** This is not a UTM projection, but rather a geographic projection), and click OK. For interval, enter 0.01 for X and 0.01 for Y. A grid will now appear on the map you are creating. Next, click on the box beside LINE STYLE. This will open the SYMBOL SELECTOR window. In this window, select LINE> SIMPLE LINE and then select NO PEN from the drop down menu next to PEN STYLE, before clicking OK. This will change the grid so that it is no longer visible. Under GRID FRAME, select EXTERIOR TICKS for FRAME STYLE. Next, click on the box next to DRAW COORDINATES. This will activate the COORDINATES options. For FORMAT, select DEGREE, MINUTE WITH SUFFIX, and then for LEFT and RIGHT change the orientation setting from HORIZONTAL to VERTICAL ASCENDING. Next, click on the box beside FONT. In the SELECT FONT window which will open, select TIMES NEW ROMAN for the FONT and 12 for the SIZE, and then click OK. Next, enter 2 for DISTANCE FROM MAP FRAME and 0 for COORDINATE PRECISION. Finally, scroll down and click on the box next to FRAME. this will add a border round the edge of your map. The contents of your PRINT COMPOSER window should now look like this:

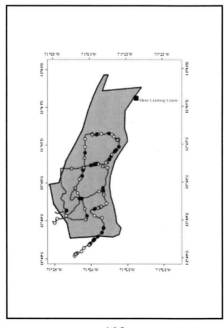

Next, you will add a scale bar. To do this, click on the LAYOUT menu on the main menu bar and select ADD SCALEBAR. Now, holding down the left hand mouse button, draw a box any where on your map to add the scale bar. At this stage, do not worry about its size or position, you will sort these out next. To do this, go back to the ITEM PROPERTIES tab, where you will notice that different options are now available. This is because the scale bar is now selected on the map (it will have an outline round it with white boxes at its corners). If you wanted to get the options for the map back, you would simply click on the map in the PRINT COMPOSER window to select it. However, for the moment, leave the scale bar selected so you can change its settings.

In the ITEM PROPERTIES tab for the scale bar, for STYLE select LINE TICKS UP. Now, scroll down to the SEGMENTS section of the tab. Here, change the options for SEGMENTS to LEFT 0 and RIGHT 4. For SIZE, enter the value 250, and for HEIGHT, enter a value of 5mm. These settings will mean that your scale bar will represent four blocks of 250m on your map. Next, scroll down to POSITION AND SIZE and click on the arrow next to it to activate the position and size options. For PAGE, enter 1 then for X enter 120, and for Y enter 225. The scale bar will now move to the bottom right hand corner of your map. Finally, scroll down and make sure that the box next to BACKGROUND is not selected, then click anywhere on your map other than on the scale bar. The contents of your PRINT COMPOSITION window should now look like this:

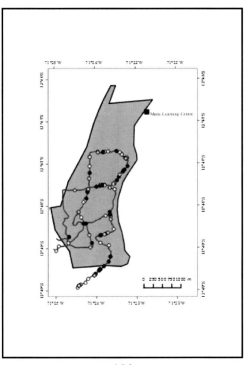

Once you have successfully inserted a scale bar, you will want to insert a North arrow to indicate which direction is north on your map. To do this, click on LAYOUT on the main menu bar and select ADD IMAGE. Now, holding down the left hand mouse button, draw a box anywhere on your map to add a blank North arrow. At this stage, do not worry about its size or position, or the fact that you cannot see the arrow, you will sort these out next. To do this, go back to the ITEM PROPERTIES tab, where you will notice that different options are now available. This is because the image you have just added is now selected on the map (it will have an outline round it with white boxes at its corners).

In the ITEM PROPERTIES tab, click on the button at the right hand end of the IMAGE SOURCE section (it has three dots in a row on it). This will open the SELECT SVG OR IMAGE FILE window. In this window, browse to C:/PROGRAM FILES/QGIS WIEN/APPS/QGIS/SVG/ARROWS (or C:/PROGRAM FILES/QGIS WIEN/APPS/QGIS-LTR/SVG/ARROWS, depending on how your computer is set up) and select NORTHARROW _04.SVG before clicking OPEN. (**NOTE:** If you are using a version of QGIS other than the version 2.8.3 recommended for this book, your QGIS folder will not be called QGIS WIEN, but will instead have the name of your version of QGIS, and you will have to navigate to that folder instead.) You will now be able to see a North arrow (or, depending on the size of the box you drew on your map, part of a North arrow) on your map. Next, for RESIZE MODE, select ZOOM, and for PLACEMENT, select MIDDLE.

NOTE: If you find that you cannot access the required SVG folder directly (for example, if you are using QGIS on a Mac OS computer), you can click on SEARCH DIRECTORY below the IMAGE SOURCE section. This will allow you to view the available image options, and you can select the appropriate North arrow image directly (see the map on the next page to find out what this looks like).

Once you have successfully selected the correct North arrow, scroll down to POSITION AND SIZE and click on the arrow next to it to activate the position and size options. For X enter 157, for Y enter 50, and then enter 10 for WIDTH and 25 for HEIGHT (**NOTE:** If you encounter any problems entering a value of 25 for HEIGHT, use the arrows at the right hand end of the HEIGHT box to change the value to 25). The North arrow will now have moved to the top right hand corner of your map. Finally, scroll down and make sure that the box next to BACKGROUND is not selected.

The contents of your PRINT COMPOSER window should now look like the image at the top of the next page.

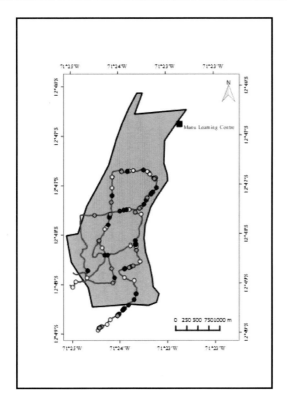

Since this is a relatively simple map, you can describe its contents in a figure legend which will tell the viewer what the different symbols and colours mean. For more complex maps, you might want to add a specific legend. This can be done through the LAYOUT menu.

All that is left now is for you to save your map and then export it. To do this, click on COMPOSER on the main menu bar and select SAVE PROJECT. Then click on COMPOSER again and select EXPORT AS IMAGE. This will open a window where you can select the format and the location where you wish to save it. The format you select will depend on what you wish to use your map for. For example, you may choose a different format depending on whether you are creating a map to include in a presentation or for a written report. For this exercise, export your map as a .jpg, call it MAP_EXERCISE_ONE and save it in the folder C:\GIS_FOR_ BIOLOGISTS. Once you have successfully completed this step, close the PRINT COMPOSER window. Finally, click on the PROJECT menu on the main menu bar on the main QGIS window and select SAVE to save the changes you have made to your GIS project.

--- Chapter Thirteen ---

Exercise Two: How To Create Your Own Feature Data Layers

In exercise one, you used existing feature data layers to create your first map, and in many instances you will be able to find such existing feature data layers to allow you to achieve your intended outcome from a specific GIS project. However, there will also be times when you will find you need to make your own data layers. This is particularly true when you have been collecting your own data, but there may also be times when you just cannot find a suitable existing data layer to work with, especially when working with fine-scale features in relatively small study areas, as will be the case in exercise two. In such instances, you will need to make your own feature data layers.

There are a number of different ways to create your own feature data layers. These include:

- **Selecting A Subset Of Features From An Existing Feature Data Layer:** This is the easiest way to create a new feature data layer and it is primarily used when you want to create a new data layer which only has some of the features from the original data layer. For example, if you have a data layer with locational records for multiple species in it, but you need a data layer which contains only those records for a single species, you could simply select the records for your species of interest and then create a new data layer based on these selected records. This is generally done within GIS software packages and you can select data based on values within fields within their attribute tables (known as 'select by attributes') or based on their spatial distribution (known as 'select by location'), as will be done in step 1 of this exercise.

106

- **Creating A New Feature Data Layer In Your GIS Software:** Most GIS software packages allow you to create new feature data layers. When you do this, you create a new empty data layer and then manually add the features you are interested in to it by drawing them in the MAP window. You can either draw the features free-hand, or your can enter sets of coordinates so that your features plot exactly where you wish them to plot.

- **Creating A Data Layer From A List Of Coordinates:** If you have a series of coordinates for specific locations in a spreadsheet or database, most GIS software packages have a tool which can be used to turn these into a point data layer. This is particularly useful for plotting the locations of sampling points or where a specific animal or species was recorded.

- **Copying A Feature From An Aerial Photograph:** If you have aerial or satellite imagery for your study area, you can use similar tools to the ones used to create new feature data layers to trace features on them and turn them into data layers. One of the most widely used ways of doing this is using a Google Earth™ image. These images provide a high quality and high resolution data set covering much of the world, and within the Google Earth interface, tools are provided which allow you to trace any features you can see. This information can then be converted into GIS data layers and used in GIS projects. This is particularly useful for generating high resolution feature data layers for relatively small study areas.

- **Importing Data Recorded On A GPS Receiver:** GPS receivers allow you to record high quality spatial information, either as waypoints (for point locations) or as tracks (for linear features). Once recorded, this information can then be converted into GIS data layers. While many GIS software packages have their own tools for doing this, there are also stand alone packages, such as DNRGPS (see page 58 for more information), which are very useful for creating GIS data layers from GPS data.

In this exercise, you will create new data layers with each of these approaches. This will be done for a study area in central Scotland, where researchers at the University of Glasgow's SCENE field station (see *tinyurl.com/GFB-Link10* for more information on this facility) are looking into factors which affect the breeding success of hole-nesting birds in a native oak woodland. You will create a map which shows where the nest boxes used in this study are located in relation to a variety of local features. This includes the shoreline of a nearby lake (called Loch Lomond), the field station where this research is conducted, the nearest road, a track which is commonly used by members of the public, and which might be a source of disturbance, and the extent of the local oak woodland. While data layers for some of these features already exist (specifically ones for the nearest road, the shoreline of the nearby lake, and for the oak woodland itself), you will need to make the data layers for the other features as part of this exercise. This will include:

1. **Nestbox_Locations:** This will be a point data layer and it will be created during this exercise from a list of coordinates in decimal degrees that give the latitude and longitude for the location of each nest box. These coordinates are in the WGS 1984 datum.

2. **SCENE_Nestbox_Locations:** This will be a point data layer and will be created during this exercise from a subset of all the nest box locations in the study. This will be done because the nest boxes have been put up in two separate patches of oak woodland, and you will only be interested in looking at those in one of these areas. This area of oak woodland is the one around the SCENE field station. This data layer will be created using the 'select by location' approach.

3. **SCENE_Buildings:** This will be a point data layer and it will be created during this exercise directly in the GIS software package you are using.

4. **Dubh_Loch:** This will be a polygon data layer which has a single polygon representing a small body of freshwater on the edge of the oak woodland around the SCENE field station. This will be created during this exercise by tracing this feature from a Google Earth image and converting it into a GIS data layer.

5. **Forestry_Track:** This will be a line data layer which represents a frequently used track through the oak woodland which could act as a source of disturbance for the hole-nesting birds of interest to this study, and so could potentially influence breeding success. The route of this track has been

mapped out by a researcher walking along it and recording their track using a GPS receiver. It has been recorded in the geographic projection using the WGS 84 datum. As part of this exercise, you will convert it into a GIS data layer.

The instructions for creating these feature data layers in ArcGIS 10.3 can be found on page 110, while those for creating the feature data layers in QGIS 2.8.3 can be found on page 133.

If you have not done so already, before you start this exercise, you will first need to create a new folder on your C: drive called GIS_FOR_BIOLOGISTS. To do this on a computer with a Windows operating system, open Windows Explorer and navigate to your C:\ drive (this may be called Windows C:). To create a new folder on this drive, right click on the window displaying the contents of your C:\ drive and select NEW> FOLDER. Now call this folder GIS_FOR_BIOLOGISTS by typing this into the folder name to replace what it is currently called (which will most likely be NEW FOLDER). This folder, which has the address C:\GIS_FOR_ BIOLOGISTS, will be used to store all files and data for the exercises in this book.

Next, you need to download the source files for three existing data layers from *www.gisinecology.com/GFB.htm#2* (**NOTE:** If you have already done this for exercise one, you do not need to download any additional data layers, as they will be in the compressed folder you downloaded at the start of that exercise). Once you have downloaded the compressed folder containing the files, make sure that you then copy all the files it contains into the folder C:\GIS_FOR_BIOLOGISTS. These existing data layers are:

1. **SCENE_Oak_Woodland.shp:** This is a polygon data layer which has a single polygon representing the area of oak woodland around the University of Glasgow's SCENE field station in central Scotland. It was created specifically for this research project. It is in the British National Grid projection and is based on the OSGB 1936 datum.

2. **SCENE_Roads.shp:** This is a line data layer which represents sections of road which run near the SCENE field station. It was obtained from the OS OpenData Meridian 2 data set (see *tinyurl.com/GFB-Link6* for more information). It is in the British National Grid projection and is based on the OSGB 1936 datum.

3. **SCENE_Loch_Lomond_Shoreline.shp:** This is a line data layer which is a high resolution representation of the shoreline of Loch Lomond. This is a large freshwater lake and the oak woodland study area stands on its shore.

This data layer was created specifically for this research project. It is in the British National Grid projection and is based on the OSGB 1936 datum.

Instructions For ArcGIS 10.3 Users:

Once you have all the required files in the correct folder on your computer, and you understand what is contained within each file, you can move on to creating your map. The starting point for this is a blank GIS project. To create a blank GIS project, first start the ArcGIS software by opening the ArcMap module. When it opens, you will be presented with a window which has the heading ARCMAP – GETTING STARTED. In this window, you can either select an existing GIS project to work on, or create a new one. To create a new blank GIS project, click on NEW MAPS in the directory tree on the left hand side and then select BLANK MAP in the right hand section of the window. Now, click OK at the bottom of this window. This will open a new blank GIS project. (**NOTE:** If this window does not appear, you can start a new project by clicking on FILE from the main menu bar area, and selecting NEW. When the NEW DOCUMENT window opens, select NEW MAPS and then BLANK MAP in the order outlined above.) Once you have opened your new GIS project, the first thing you need to do is save it under a new, and meaningful, name. To do this, click on FILE on the main menu bar area, and select SAVE AS. Save it as EXERCISE_TWO in the folder C:\GIS_FOR_BIOLOGISTS.

STEP 1: SET THE PROJECTION AND COORDINATE SYSTEM OF YOUR DATA FRAME:

Whenever you start a new GIS project, the first thing you should do, even before you add any data layers to it, is select an appropriate projection/coordinate system and then set the data frame (i.e. the map you are going to add your data layers to) to use this projection/coordinate system. This ensures that you do not accidently end up using one which is inappropriate. In addition, most problems that beginners have with GIS are caused by a failure to use an appropriate projection/coordinate system or a conflict between the projection/coordinate system used for data layers and the one being used for the data frame itself. For this exercise, you will use a pre-existing projection/coordinate system called British National Grid. This is a transverse mercator projection which is specifically designed to minimise the distortion of features in the British Isles and it uses the OSGB 1936 datum.

To set your data frame to use the British National Grid projection/coordinate system, work through the following flow diagram:

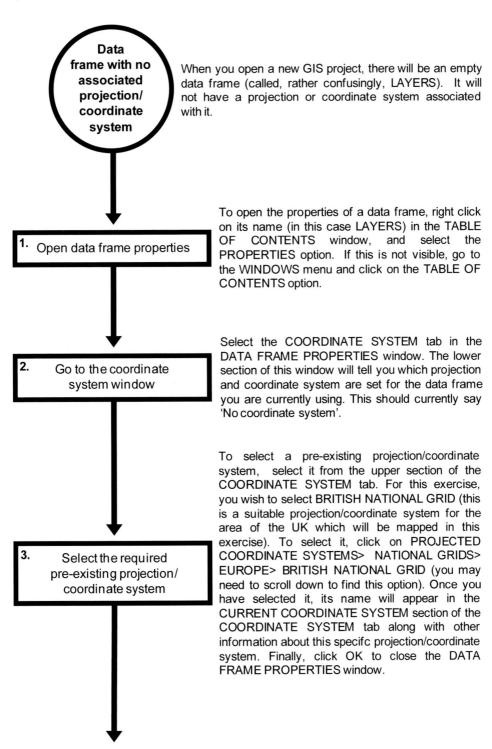

Data frame with no associated projection/coordinate system

When you open a new GIS project, there will be an empty data frame (called, rather confusingly, LAYERS). It will not have a projection or coordinate system associated with it.

1. Open data frame properties

To open the properties of a data frame, right click on its name (in this case LAYERS) in the TABLE OF CONTENTS window, and select the PROPERTIES option. If this is not visible, go to the WINDOWS menu and click on the TABLE OF CONTENTS option.

2. Go to the coordinate system window

Select the COORDINATE SYSTEM tab in the DATA FRAME PROPERTIES window. The lower section of this window will tell you which projection and coordinate system are set for the data frame you are currently using. This should currently say 'No coordinate system'.

3. Select the required pre-existing projection/coordinate system

To select a pre-existing projection/coordinate system, select it from the upper section of the COORDINATE SYSTEM tab. For this exercise, you wish to select BRITISH NATIONAL GRID (this is a suitable projection/coordinate system for the area of the UK which will be mapped in this exercise). To select it, click on PROJECTED COORDINATE SYSTEMS> NATIONAL GRIDS> EUROPE> BRITISH NATIONAL GRID (you may need to scroll down to find this option). Once you have selected it, its name will appear in the CURRENT COORDINATE SYSTEM section of the COORDINATE SYSTEM tab along with other information about this specifc projection/coordinate system. Finally, click OK to close the DATA FRAME PROPERTIES window.

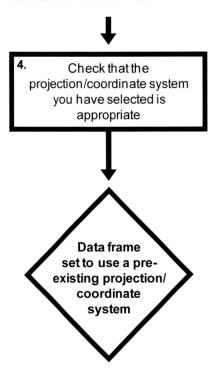

4. Check that the projection/coordinate system you have selected is appropriate

Once you have selected your chosen projection/ coordinate system, you need to check that it is appropriate. This involves examining how data layers look in it. For this exercise, this will be done in the next step by adding a series of data layers and checking that the features in them plot in the expected places and that any polygons have the expected shape.

Data frame set to use a pre-existing projection/ coordinate system

To check that you have done this step properly, right click on the name of your data frame (LAYERS) in the TABLE OF CONTENTS window and select PROPERTIES. Click on the COORDINATE SYSTEM tab of the DATA FRAME PROPERTIES window and make sure that the contents of the CURRENT COORDINATE SYSTEM section of the window has the following text at the top of it:

> British_National_Grid
> WKID: 27700 Authority EPSG
>
> Projection: Transverse_Mercator
> False_Easting: 400000.0
> False_Northing: -100000.0
> Central_Meridian: -2
> Scale_Factor: 0.9996012717
> Latitude_Of_Origin: 49.0
> Linear Unit: Meter (1.0)

If it does not, you will need to repeat this step until you have assigned the correct projection/coordinate system to your data frame. Once you have successfully completed this step, click on the FILE menu on the main menu bar and select SAVE to save the changes you have made to your GIS project.

STEP 2: ADD ANY EXISTING DATA LAYERS YOU WISH TO DISPLAY ON YOUR MAP TO YOUR GIS PROJECT:

Once you have set the projection/coordinate system of your data frame, you are ready to add some existing data layers to it. For this exercise, you will use a number of existing data layers which provide information about a variety of different features for the area of the native oak woodland study area. This will include the area of the woodland itself (SCENE_OAK_WOODLAND), information about the position of the shoreline of a large lake (LOCH_LOMOND_SHORELINE) and information about nearby roads (SCENE_ROADS). These existing data layers are all in the British National Grid projection/coordinate system which is being used for this exercise.

To add these data layers to your GIS project, work through the following flow diagram:

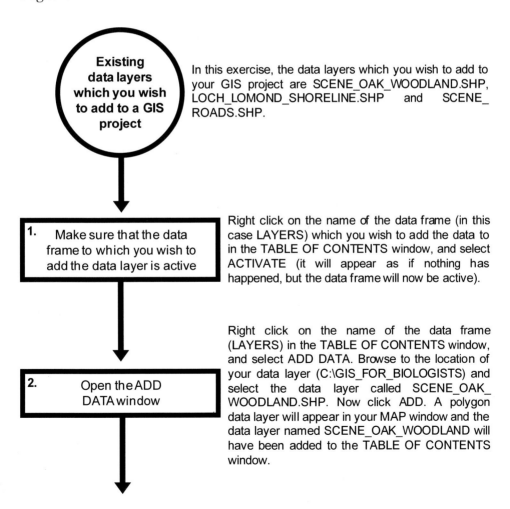

Existing data layers which you wish to add to a GIS project

In this exercise, the data layers which you wish to add to your GIS project are SCENE_OAK_WOODLAND.SHP, LOCH_LOMOND_SHORELINE.SHP and SCENE_ROADS.SHP.

1. Make sure that the data frame to which you wish to add the data layer is active

Right click on the name of the data frame (in this case LAYERS) which you wish to add the data to in the TABLE OF CONTENTS window, and select ACTIVATE (it will appear as if nothing has happened, but the data frame will now be active).

2. Open the ADD DATA window

Right click on the name of the data frame (LAYERS) in the TABLE OF CONTENTS window, and select ADD DATA. Browse to the location of your data layer (C:\GIS_FOR_BIOLOGISTS) and select the data layer called SCENE_OAK_WOODLAND.SHP. Now click ADD. A polygon data layer will appear in your MAP window and the data layer named SCENE_OAK_WOODLAND will have been added to the TABLE OF CONTENTS window.

3. Check the projection/ coordinate system for the newly added data layer

Whenever you add a data layer to a GIS project, you should always check that is has a projection/coordinate system assigned to it, and look at what this projection/coordinate system is. This is so that you know whether you will need to assign a projection/coordinate system to it, or transform it into a different projection/coordinate system before you can use it in your GIS project. To check the projection/coordinate system of your newly added data layer, right click on its name in the TABLE OF CONTENTS window and select PROPERTIES. In the LAYER PROPERTIES window which opens, click on the SOURCE tab and check that the projection/coordinate system listed under DATA SOURCE is BRITISH_ NATIONAL_GRID (the same as the data frame). Now click OK to close the LAYER PROPERTIES window. If the data layer had not been in the same projection/coordinate system as the data frame, you would have to have used the PROJECT tool to transform it into the correct projection/coordinate system. You wil learn how to do this later in this exercise.

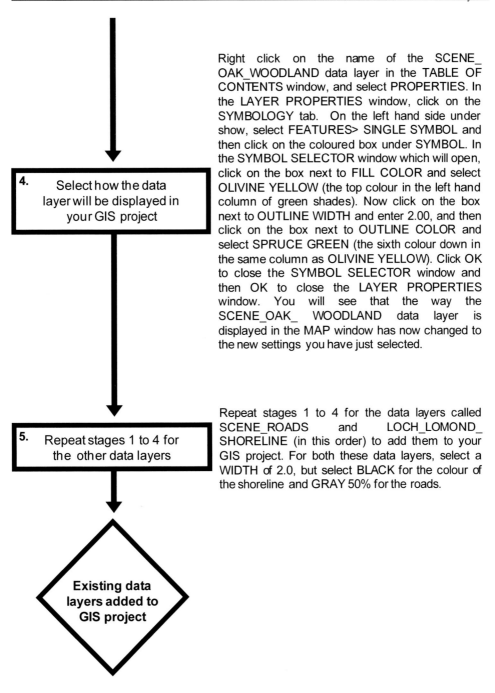

4. Select how the data layer will be displayed in your GIS project

Right click on the name of the SCENE_ OAK_WOODLAND data layer in the TABLE OF CONTENTS window, and select PROPERTIES. In the LAYER PROPERTIES window, click on the SYMBOLOGY tab. On the left hand side under show, select FEATURES> SINGLE SYMBOL and then click on the coloured box under SYMBOL. In the SYMBOL SELECTOR window which will open, click on the box next to FILL COLOR and select OLIVINE YELLOW (the top colour in the left hand column of green shades). Now click on the box next to OUTLINE WIDTH and enter 2.00, and then click on the box next to OUTLINE COLOR and select SPRUCE GREEN (the sixth colour down in the same column as OLIVINE YELLOW). Click OK to close the SYMBOL SELECTOR window and then OK to close the LAYER PROPERTIES window. You will see that the way the SCENE_OAK_ WOODLAND data layer is displayed in the MAP window has now changed to the new settings you have just selected.

5. Repeat stages 1 to 4 for the other data layers

Repeat stages 1 to 4 for the data layers called SCENE_ROADS and LOCH_LOMOND_ SHORELINE (in this order) to add them to your GIS project. For both these data layers, select a WIDTH of 2.0, but select BLACK for the colour of the shoreline and GRAY 50% for the roads.

Existing data layers added to GIS project

At the end of this step, your TABLE OF CONTENTS window should look like the image at the top of the next page.

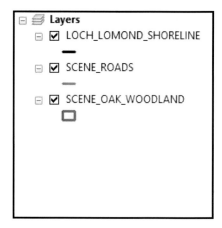

While the contents of your MAP window should look like this:

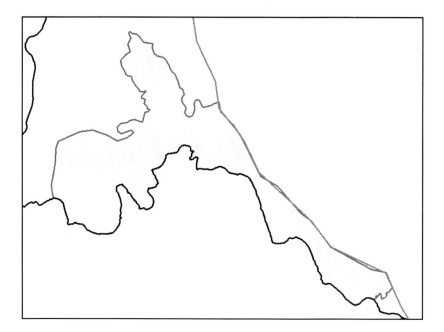

If it does not look like this, try right-clicking on the name of the SCENE_OAK_WOODLAND data layer in the TABLE OF CONTENTS window and selecting ZOOM TO LAYER. If it still does not look right, remove all the data layers from your GIS project by right-clicking on their names in the TABLE OF CONTENTS window and selecting REMOVE. Next, go back to step one and ensure that you have set the projection/coordinate system of your data frame correctly, and then repeat step two. Once you have successfully completed this step, click on the FILE menu on the main menu bar and select SAVE to save the changes you have made to your GIS project.

STEP 3: CREATE A NEW POINT DATA LAYER FROM A LIST OF COORDINATES IN A SPREADSHEET OR DATABASE:

Now that you have added the required existing data layers to your GIS project, you can start creating new ones to add further information to your map. The first new data layer you will create will be a point data layer of the nest box locations used to study the breeding success of hole-nesting birds in the oak woodland from a list of coordinates held in a spreadsheet. This will be done in two parts. Firstly, you will plot the locations of all the nest boxes in the spreadsheet and make them into a point data layer. Once you have done this, you will select only those nest boxes which are within the oak woodland study area around the SCENE field station and make a new data layer with just these locations in it. Therefore, in this step, you will learn two different ways of making a new data layer from existing data sets.

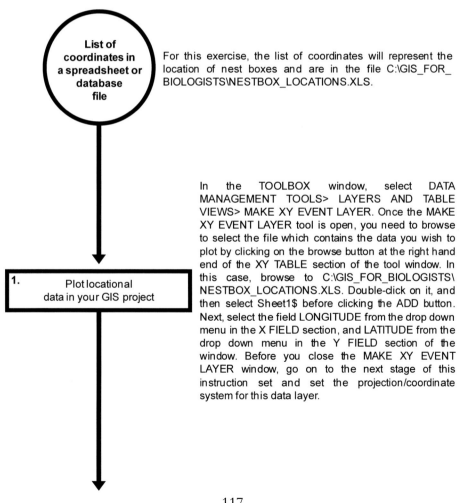

List of coordinates in a spreadsheet or database file

For this exercise, the list of coordinates will represent the location of nest boxes and are in the file C:\GIS_FOR_ BIOLOGISTS\NESTBOX_LOCATIONS.XLS.

1. Plot locational data in your GIS project

In the TOOLBOX window, select DATA MANAGEMENT TOOLS> LAYERS AND TABLE VIEWS> MAKE XY EVENT LAYER. Once the MAKE XY EVENT LAYER tool is open, you need to browse to select the file which contains the data you wish to plot by clicking on the browse button at the right hand end of the XY TABLE section of the tool window. In this case, browse to C:\GIS_FOR_BIOLOGISTS\ NESTBOX_LOCATIONS.XLS. Double-click on it, and then select Sheet1$ before clicking the ADD button. Next, select the field LONGITUDE from the drop down menu in the X FIELD section, and LATITUDE from the drop down menu in the Y FIELD section of the window. Before you close the MAKE XY EVENT LAYER window, go on to the next stage of this instruction set and set the projection/coordinate system for this data layer.

2. Assign the correct projection/coordinate system to the data layer

In the MAKE XY EVENT LAYER window, click on the button at the right hand end of the SPATIAL REFERENCES (OPTIONAL) section. This will open the SPATIAL REFERENCE PROPERTIES window. In this window, select GEOGRAPHIC COORDINATE SYSTEM> WORLD> WGS 1984, then click on the OK button. You are selecting this projection/coordinate system because you are using latitude and longitude values with the WGS 1984 datum to plot your species locations. **NOTE:** If your coordinates were in a different projection/coordinate system, you would set this different system here. Next, click on OK in the SPATIAL REFERENCE PROPERTIES window. The SPATIAL REFERENCES (OPTIONAL) section should now contain the text GCS_WGS_1984. Finally, click on OK to close the MAKE XY EVENT LAYER window. **NOTE:** Do not worry if you get a warning message about a datum conflict. You will sort this out next.

3. Transform your data layer into the projection/coordinate system being used for your GIS project and save it as a shapefile

At the moment, the data layer that you have made from the spreadsheet (SHEET1$_LAYER) is not a permanent layer, nor is it in the projection/coordinate system being used for this GIS project. Therefore, the next step is to convert it into a permanent shapefile that is in the right projection/coordination system. This is done with the PROJECT tool. To open this tool, in the TOOLBOX window, select DATA MANAGEMENT TOOLS> PROJECTIONS AND TRANSFORMATIONS> PROJECT (**NOTE:** In older versions of ArcGIS, there will be an additional FEATURES folder that you will need to click on to find the PROJECT tool). In the PROJECT window, select SHEET1$_LAYER from the drop down menu under INPUT DATASET OR FEATURE CLASS. Next, type C:\GIS_FOR_ BIOLOGISTS\NESTBOX_LOCATIONS.SHP in the OUTPUT DATASET OR FEATURE CLASS section. Finally, in the OUTPUT COORDINATE SYSTEM section, click on the button at the right hand end to open the SPATIAL REFERENCE PROPERTIES window. In the window which opens, select PROJECTED COORDINATE SYSTEMS> NATIONAL GRIDS> EUROPE> BRITISH NATIONAL GRID and then click OK. Now, click OK in the PROJECT tool window to create the projected data layer. Finally, right-click on the temporary data layer called SHEET1$_LAYER and select REMOVE. **NOTE:** If the data layer you have just made (NESTBOX_ LOCATIONS) is not automatically added to your GIS PROJECT, you will need to add it manually. To do this, right click on the name of the data frame (LAYERS) in the TABLE OF CONTENTS window, and select ADD DATA, then add it from the folder C:\GIS_FOR_ BIOLOGISTS in the same way you have added all previous data layers.

118

If you right click on the name of the data layer you have just created (NESTBOX_LOCATIONS) in the TABLE OF CONTENTS window and select ZOOM TO LAYER, you will see that there are two groups of nest boxes in it, one set in the SCENE_OAK_ WOODLAND polygon and one further south. You now want to select those nest boxes in the SCENE oak woodland and make a new data layer with just these locations in it. To do this, click on SELECTION on the main menu bar and select SELECT BY LOCATION. This will open the SELECT BY LOCATION window. For SELECTION METHOD, choose SELECT FEATURES FROM and then in the TARGET LAYER(S) section, select the data layer NESTBOX_LOCATIONS. Next, in SOURCE LAYER section, select SCENE_OAK_WOODLAND and then for SPATIAL SELECTION METHOD FOR TARGET LAYER FEATURE(S), select ARE WITHIN THE SOURCE LAYER FEATURE. Now click OK to run the SELECTION tool. Once it has finished, you will see that in the MAP window, only those nest boxes which are in the SCENE oak woodland have been selected. To make a new data layer containing only these locations, right click on the data layer called NESTBOX_LOCATIONS in the TABLE OF CONTENTS window and select DATA> EXPORT DATA. In the EXPORT DATA window which opens, beside EXPORT make sure you select SELECTED FEATURES from the options available in the drop down menu. Finally, type C:\GIS_FOR_BIOLOGISTS\ SCENE_NESTBOX_LOCATIONS.SHP into the OUTPUT FEATURE CLASS section and click OK to make your new data layer with just the selected features in it. If asked, click YES to add this new data layer to your GIS project.

Right click on the name of the SCENE_NESTBOX_LOCATIONS data layer in the TABLE OF CONTENTS window, and select PROPERTIES. In the LAYER PROPERTIES window, click on the SYMBOLOGY tab. On the left hand side, under SHOW, select FEATURES> SINGLE SYMBOL. Next, under SYMBOL in the top middle of the LAYER PROPERTIES window, click on the button with the coloured circle on it. This will open the SYMBOL SELECTOR window. Select CIRCLE 2 for the symbol by clicking on it in the left hand section of the window. Then, on the right hand side of the SYMBOL SELECTOR window, click on the button next to COLOR and hover the cursor over the fourth red box down. This will tell you that this colour's name is PONSETTIA RED. Select this by clicking on it. Next, select 12.0 for SIZE. Finally, click OK to close the SYMBOL SELECTOR window and then OK to close the LAYER PROPERTIES window.

4. Make a new data layer by selecting a subset of data from an existing data layer

5. Select how your new point data layer will be displayed

Point data layer created from a list of coordinates

119

At the end of this step, your TABLE OF CONTENTS window should look like this:

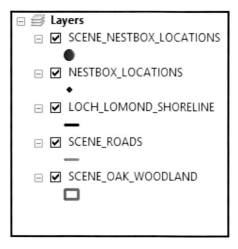

Click on the box next to the name of the data layer SCENE_NESTBOX_ LOCATIONS in the TABLE OF CONTENTS window so that it is no longer displayed in the MAP window. Now click on SELECTION on the main menu bar and select CLEAR SELECTED FEATURES. The contents of your MAP window should now look like this:

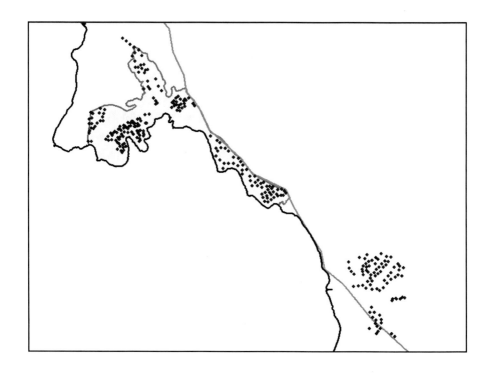

Check the box next to SCENE_NESTBOX_LOCATIONS again (so that it is displayed in the MAP window again) and uncheck the box next to NESTBOX_LOCATIONS (so that it is no longer displayed in the MAP window). You will now see that it contains only those nest box locations in the SCENE oak woodland. Right click on SCENE_OAK_WOODLAND in the TABLE OF CONTENTS window and select ZOOM TO LAYER. The contents of your MAP window should now look like this:

Once you have successfully completed this step, click on the FILE menu on the main menu bar and select SAVE to save the changes you have made to your GIS project.

STEP 4: CREATE A NEW DATA LAYER WITHIN YOUR GIS SOFTWARE:

In step three, you learned how to make new data layers in two different ways. In this step, you will learn how to make a new data layer in a third way. This is directly in the MAP window of your GIS software. You can also make polygon and line data layers in this way, but the data layer you will create in this step will be a point data layer. This data layer will have a single point marking the location of the University of Glasgow's SCENE field station buildings and it will be called SCENE_BUILDINGS. This data layer can be made by working through the flow diagram which starts on the next page.

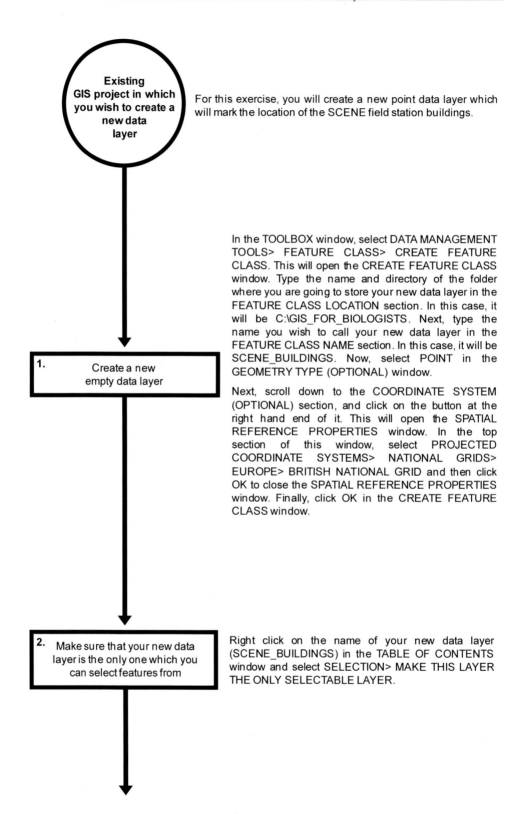

Existing GIS project in which you wish to create a new data layer

For this exercise, you will create a new point data layer which will mark the location of the SCENE field station buildings.

1. Create a new empty data layer

In the TOOLBOX window, select DATA MANAGEMENT TOOLS> FEATURE CLASS> CREATE FEATURE CLASS. This will open the CREATE FEATURE CLASS window. Type the name and directory of the folder where you are going to store your new data layer in the FEATURE CLASS LOCATION section. In this case, it will be C:\GIS_FOR_BIOLOGISTS. Next, type the name you wish to call your new data layer in the FEATURE CLASS NAME section. In this case, it will be SCENE_BUILDINGS. Now, select POINT in the GEOMETRY TYPE (OPTIONAL) window.

Next, scroll down to the COORDINATE SYSTEM (OPTIONAL) section, and click on the button at the right hand end of it. This will open the SPATIAL REFERENCE PROPERTIES window. In the top section of this window, select PROJECTED COORDINATE SYSTEMS> NATIONAL GRIDS> EUROPE> BRITISH NATIONAL GRID and then click OK to close the SPATIAL REFERENCE PROPERTIES window. Finally, click OK in the CREATE FEATURE CLASS window.

2. Make sure that your new data layer is the only one which you can select features from

Right click on the name of your new data layer (SCENE_BUILDINGS) in the TABLE OF CONTENTS window and select SELECTION> MAKE THIS LAYER THE ONLY SELECTABLE LAYER.

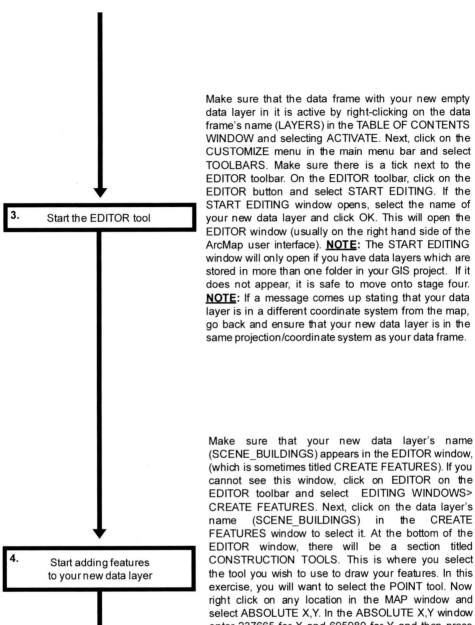

3. Start the EDITOR tool

Make sure that the data frame with your new empty data layer in it is active by right-clicking on the data frame's name (LAYERS) in the TABLE OF CONTENTS WINDOW and selecting ACTIVATE. Next, click on the CUSTOMIZE menu in the main menu bar and select TOOLBARS. Make sure there is a tick next to the EDITOR toolbar. On the EDITOR toolbar, click on the EDITOR button and select START EDITING. If the START EDITING window opens, select the name of your new data layer and click OK. This will open the EDITOR window (usually on the right hand side of the ArcMap user interface). **NOTE:** The START EDITING window will only open if you have data layers which are stored in more than one folder in your GIS project. If it does not appear, it is safe to move onto stage four. **NOTE:** If a message comes up stating that your data layer is in a different coordinate system from the map, go back and ensure that your new data layer is in the same projection/coordinate system as your data frame.

4. Start adding features to your new data layer

Make sure that your new data layer's name (SCENE_BUILDINGS) appears in the EDITOR window, (which is sometimes titled CREATE FEATURES). If you cannot see this window, click on EDITOR on the EDITOR toolbar and select EDITING WINDOWS> CREATE FEATURES. Next, click on the data layer's name (SCENE_BUILDINGS) in the CREATE FEATURES window to select it. At the bottom of the EDITOR window, there will be a section titled CONSTRUCTION TOOLS. This is where you select the tool you wish to use to draw your features. In this exercise, you will want to select the POINT tool. Now right click on any location in the MAP window and select ABSOLUTE X,Y. In the ABSOLUTE X,Y window enter 237665 for X and 695980 for Y and then press the ENTER key on your keyboard. This will create a new point at the location of the coordinates which you entered. **NOTE:** You can also simply click on the MAP window with the POINT tool to add a new point at a location of your choosing.

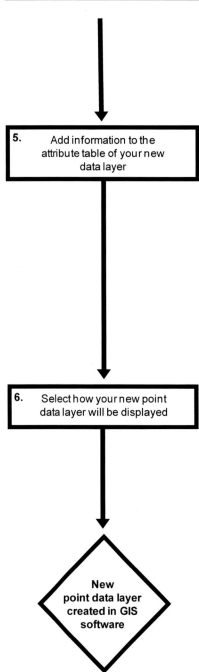

5. Add information to the attribute table of your new data layer

6. Select how your new point data layer will be displayed

New point data layer created in GIS software

On the EDITOR toolbar, click on EDITOR and select STOP EDITING. When asked, save your edits. Next, right click on the data layer's name in the TABLE OF CONTENTS window and select OPEN ATTRIBUTE TABLE. Click on the TABLE OPTIONS button (in the top left hand corner of the TABLE window) and select ADD FIELD. In the ADD FIELD window, enter LOCATION in the NAME section. For TYPE, select TEXT and enter 50 for LENGTH before clicking OK to create the new field. Once the new field has been created, move the TABLE window so you can see the MAP window beneath it, then go back to the EDITOR toolbar and select EDITOR> START EDITING again. You can now click on individual cells in the attribute table and fill them in. For LOCATION, enter the following text: SCENE Buildings. Finally, go back to the EDITOR tool bar in the MAP window, and select EDITOR> STOP EDITING. When asked, save your edits.

Right click on the name of the SCENE_BUILDINGS data layer in the TABLE OF CONTENTS window, and select PROPERTIES. In the LAYER PROPERTIES window, click on the SYMBOLOGY tab. On the left hand side, under SHOW, select FEATURES> SINGLE SYMBOL. Next, under SYMBOL in the top middle of the LAYER PROPERTIES window, click on the button with the coloured circle on it. This will open the SYMBOL SELECTOR window. Select SQUARE for the symbol by clicking on it in the left hand section of the window, then, on the right hand side of the SYMBOL SELECTOR window, select 14.0 for size. Finally, click OK to close the SYMBOL SELECTOR window and then click OK to close the LAYER PROPERTIES window.

At the end of this step, the contents of your TABLE window should look like this:

FID	Shape *	Id	Location
0	Point	0	SCENE Buildings

Now close the TABLE window and then the CREATE FEATURES window. Your TABLE OF CONTENTS window should now look like this:

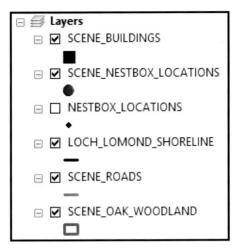

While the contents of your MAP window should look like this:

Once you have successfully completed this step, click on the FILE menu on the main menu bar and select SAVE to save the changes you have made to your GIS project.

125

STEP 5: CREATE A NEW POLYGON DATA LAYER BY TRACING A FEATURE FROM A GOOGLE EARTH IMAGE:

Google Earth images provide high quality aerial images covering much of the world, and this means that they are a great source of information which biologists can use for many different purposes. In this step, you will learn how to create a polygon data layer by tracing a feature in a Google Earth image and then importing it into your GIS project from the Google Earth data layer format (which has the extension .kml or .kmz). The feature you will trace is a small lake, called the Dubh Loch, which borders on the oak woodland study area. **NOTE:** To be able to do this step, you will need to have a copy of the Google Earth user interface installed on your computer, and you will need to have access to the internet. If you do not have either of these, you can skip this step and move on to step 6. The polygon data layer of the Dubh Loch can be made by working through the following flow diagram:

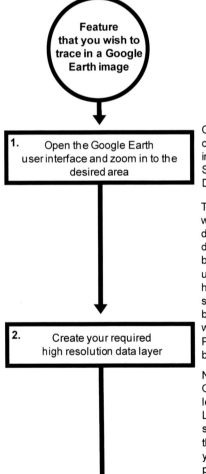

Open the Google Earth user interface. Once it opens, type DUBH LOCHAN, GLASGOW, G63, UK into the SEARCH window and then click the SEARCH button. This will allow you to locate the Dubh Loch.

To create a high resolution data layer, first decide whether you want to create a point data layer, a line data layer or a polygon data layer. Once you have decided this, select the appropriate tool from the buttons on the main menu bar of the Google Earth user interface. For this exercise, you will make a high resolution polygon data layer of the Dubh Loch, so you will need to select the ADD POLYGON tool by clicking on ADD and selecting POLYGON. This will open a window called GOOGLE EARTH – NEW POLYGON. In this window, type DUBH LOCH beside NAME.

Now move this window so that you can see all of the Google Earth MAP window. Starting at the bottom left hand corner, trace out the outline of the Dubh Loch by clicking at regular intervals along its shoreline. If you find that you have added a point in the wrong place, simply press the DELETE key on your keyboard. When you are happy with the polygon you have drawn, click OK to close the GOOGLE EARTH – NEW POLYGON window.

126

Under PLACES on the left hand side of the Google Earth user interface, right click on the name DUBH LOCH and select SAVE PLACE AS. This will open the SAVE FILE window. **NOTE:** You may have to scroll down to find this data layer. Browse to the folder C:\GIS_FOR BIOLOGISTS and then type in the name DUBH_LOCH. Next, under SAVE AS TYPE select KML (*.KML) and then click the SAVE button. You can now close the Google Earth user interface.

Next, return to the ArcMAP module of the ArcGIS software package. In the TOOLBOX window, select CONVERSION TOOLS> FROM KML> KML TO LAYER. In the INPUT KML section of the window which opens, click on the browse button at the right hand end. In the OPEN window which will appear, select the file C:\GIS_FOR_BIOLOGISTS\ DUBH_LOCH.KML and then click OPEN. For OUTPUT LOCATION, type C:\GIS_FOR_ BIOLOGISTS\DUBH_LOCH_LAYER. Now click OK to create a layer file from your Google Earth data layer. Once the layer has been created, you will receive a warning telling you that there is a datum conflict. This is because Google Earth uses the WGS 1984 datum and your map is using the OSGB 1936 datum. Do not worry about this, you will sort this out next.

3. Convert your Google Earth data layer into a shapefile

In the TOOLBOX window, select DATA MANAGEMENT TOOLS> PROJECTIONS AND TRANSFORMATIONS> PROJECT. In the PROJECT window which opens, select DUBH_ LOCH> POLYGONS from the drop down menu for INPUT DATASET OR FEATURE CLASS. Next, for OUTPUT DATA SET OR FEATURE CLASS, enter C:\GIS_FOR_BIOLOGISTS\DUBH_ LOCH.SHP. Now, click on the button at the right hand end of the OUTPUT COORDINATE SYSTEM window and in the SPATIAL REFERENCE PROPERTIES window which opens, select PROJECTED COORDINATE SYSTEMS> NATIONAL GRIDS> EUROPE> BRITISH NATIONAL GRID, and then click OK. Finally, click OK to run the PROJECT tool. **NOTE:** If the data layer you have just made is not automatically added to your GIS project, you will need to add it manually by right-clicking on the name of your data frame (LAYERS) in the TABLE OF CONTENTS window and selecting ADD DATA. In the ADD DATA window, browse to the folder C:\GIS_FOR_BIOLOGISTS and select the data layer DUBH_LOCH.SHP. Once the newly created DUBH_LOCH data layer has been added to your GIS project, you can remove the data layer called DUBH_LOCH_LAYER from it. To do this, right click on its name in the TABLE OF CONTENTS window and select REMOVE.

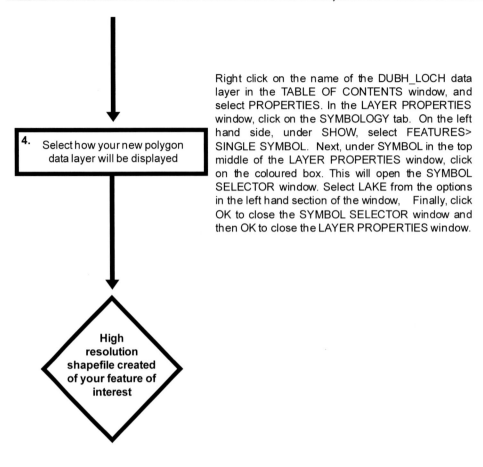

4. Select how your new polygon data layer will be displayed

Right click on the name of the DUBH_LOCH data layer in the TABLE OF CONTENTS window, and select PROPERTIES. In the LAYER PROPERTIES window, click on the SYMBOLOGY tab. On the left hand side, under SHOW, select FEATURES> SINGLE SYMBOL. Next, under SYMBOL in the top middle of the LAYER PROPERTIES window, click on the coloured box. This will open the SYMBOL SELECTOR window. Select LAKE from the options in the left hand section of the window,   Finally, click OK to close the SYMBOL SELECTOR window and then OK to close the LAYER PROPERTIES window.

High resolution shapefile created of your feature of interest

At the end of this step, your TABLE OF CONTENTS window should look like this:

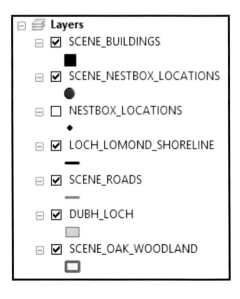

While the contents of your MAP window should look like this:

Once you have successfully completed this step, click on the FILE menu on the main menu bar and select SAVE to save the changes you have made to your GIS project.

STEP 6: CREATE A NEW LINE DATA LAYER BY IMPORTING A TRACK RECORDED ON A GPS RECEIVER:

The final way you will create a new data layer in this exercise is from data recorded on a GPS receiver. GPS receivers can record two types of data: waypoints, which mark specific locations, and tracks, which are lines marking specific routes. You can use such tracks to record features which you might want to add to a GIS project simply by setting up your GPS to record a track and then walking along the feature of interest at a steady pace. Once you reach the end of the feature, you can save the section of track you have just recorded and then load it into a GIS project. In this step, you will do this for a GPS track file (which, like all GPS-compatible files, has the extension .gpx) that was recorded by walking along a forestry track which runs through the middle of the oak woodland study area. This will allow you to add this feature to the map you are building of the oak woodland study area. The line data layer of the forestry track can be created by working through the flow diagram which starts on the next page. To find out more about how to set up a GPS to record a track, visit *www.GISinEcology.com/GFB.htm#video9*.

GPS data file that you want to convert into a GIS-compatible format

1. Convert your GPS file into the shapefile format

2. Transform the projection/ coordinate system of your data layer into the one you are using for your GIS project

In the TOOLBOX window, select CONVERSION TOOLS> FROM GPS> GPX TO FEATURES. For INPUT GPX FILE, click on the browse button at the right hand end. Browse to the folder C:\GIS_ FOR_BIOLOGISTS, select the file FORESTRY_ TRACK.GPX and then click OPEN. In the OUTPUT FEATURE CLASS section of the GPX TO FEATURES window, type C:\GIS_FOR_BIOLOGISTS\FORESTRY_ TRACK_V1.SHP. Finally, click OK to run this tool. Do not worry if you get a warning about a conflict between datums, you will sort this out next.

You will have a data layer called FORESTRY_TRACK_ V1 added to your GIS project. However, it is in a different projection/coordinate system (the geographic projection based on the WGS 1984 datum) than the rest of your GIS project (British National Grid). This means that before you can use it in your GIS project, you need to transform it into the same projection/coordinate system as your data frame and your other data layers. This is done by opening the PROJECT tool.

In the TOOLBOX window, select DATA MANAGEMENT TOOLS> PROJECTIONS AND TRANSFORMATIONS> PROJECT. In the PROJECT window which opens, select FORESTRY_TRACK_V1 from the drop down menu for INPUT DATASET OR FEATURE CLASS. Next, for OUTPUT DATA SET OR FEATURE CLASS, type C:\GIS_FOR_BIOLOGISTS\FORESTRY_ TRACK_V2.SHP. Now, click on the button at the right hand end of the OUTPUT COORDINATE SYSTEM window. In the SPATIAL REFERENCE PROPERTIES window which will open, select PROJECTED COORDINATE SYSTEMS> NATIONAL GRIDS> EUROPE> BRITISH NATIONAL GRID, and then click OK. Finally, click OK to run the PROJECT tool. **NOTE:** If the data layer you have just made is not automatically added to your GIS project, you will need to add it manually by right-clicking on the name of your data frame (LAYERS) in the TABLE OF CONTENTS window and selecting ADD DATA. In the ADD DATA window, browse to the folder C:/GIS_FOR_ BIOLOGISTS and select the data layer FORESTRY_ TRACK_V2.SHP. Once your new data layer has been added to your GIS project, you can remove the data layer FORESTRY_TRACK_V1 from it. To do this, right click on its name in the TABLE OF CONTENTS window and select REMOVE.

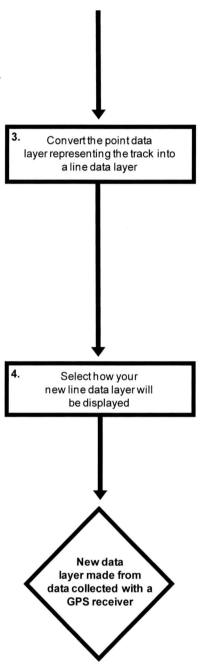

3. Convert the point data layer representing the track into a line data layer

If you examine the data layer you have just made (FORESTRY_TRACK_V2) in the MAP window, you will see that it is represented by a series of points and not a line. In order to recreate the line representing the forestry track recorded by the GPS, you will need to create a line joining all these points together. To do this, select DATA MANAGEMENT TOOLS> FEATURES> POINTS TO LINE in the TOOLBOX window. In the window that opens, select FORESTRY_TRACK_V2 from the drop down menu for INPUT FEATURES, and then type C:\GIS_FOR_BIOLOGISTS\FORESTY_ TRACK_V3 into the OUTPUT FIELD section and click on OK to run the tool. Once the tool has finished running, remove the data layer called FORESTRY_ TRACK_V2 from your GIS project by right-clicking on its name in the TABLE OF CONTENTS window and selecting REMOVE.

4. Select how your new line data layer will be displayed

Right click on the name of the FORESTRY_TRACK_V3 data layer in the TABLE OF CONTENTS window, and select PROPERTIES. In the LAYER PROPERTIES window, click on the SYMBOLOGY tab. On the left hand side under SHOW, select FEATURES> SINGLE SYMBOL. Next, under SYMBOL in the top middle of the LAYER PROPERTIES window, click on the box with the line in it. This will open the SYMBOL SELECTOR window. Click on the box next to COLOR and select GRAY 20%. Next enter 2.00 for WIDTH. Finally, click OK to close the SYMBOL SELECTOR window and then OK to close the LAYER PROPERTIES window.

New data layer made from data collected with a GPS receiver

At the end of this step, your TABLE OF CONTENTS window should look like the image at the top of the next page.

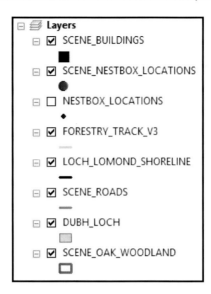

While the contents of your MAP window should look like this:

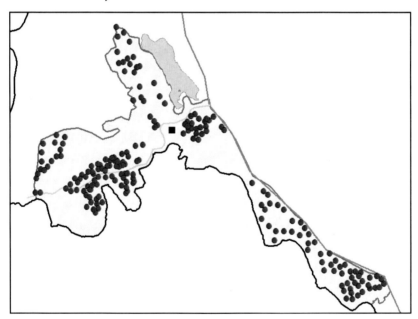

Once you have successfully completed this step, click on the FILE menu on the main menu bar and select SAVE to save the changes you have made to your GIS project.

This is the end of exercise two. Now that you have completed it, you know all the main ways that you can make new feature data layers in a GIS project. To recap, these are: 1. Creating a point data layer from a list of coordinates stored in a spreadsheet or database; 2. By selecting a subset of data in an existing data layer; 3. Creating a new data layer directly in your GIS software; 4. Creating a data layer by tracing a feature in a Google Earth image; 5. Creating a data layer using a track or waypoint file recorded on a GPS receiver.

Instructions For QGIS 2.8.3 Users:

Once you have the required files downloaded into the correct folder on your computer, and you understand what is contained within each file, you can move onto creating your map. The starting point for this is a blank GIS project. To create a blank GIS project, first open QGIS. Once it is open, click on the PROJECT menu and select SAVE AS. In the window which opens, save your GIS project as EXERCISE_TWO in the folder C:\GIS_FOR_ BIOLOGISTS. **NOTE:** Before you start this exercise, you need to ensure that the SPATIAL QUERY plugin is installed and activated in your version of QGIS. To do this, click on PLUGINS on the main menu bar, and select MANAGE AND INSTALL PLUGINS (you will need to have access to the internet to be able to manage your plugins using this option). In the PLUGINS window which will open, click on the ALL tab on the left hand side and then type SPATIAL QUERY into the SEARCH section of this window. If the SPATIAL QUERY PLUGIN is not installed, select it and click on the INSTALL button. If it is installed, check that there is a cross (or a tick) in the white box next to its name (this means that it has been activated). If there is no cross or tick in this box, click on it so that a cross/tick appears in it. You can now click CLOSE to close the PLUGINS window.

STEP 1: SET THE PROJECTION AND COORDINATE SYSTEM OF YOUR DATA FRAME:

Whenever you start a new GIS project, the first thing you should do, even before you add any data layers to it, is select an appropriate projection/coordinate system and then set the data frame (i.e. the map you are going to add your data layers to) to use this projection/coordinate system. This ensures that you do not accidently end up using one which is inappropriate. In addition, most problems that beginners have with GIS are caused by a failure to use an appropriate projection/coordinate system or a conflict between the projection/coordinate system used for data layers and the one being used for the data frame itself. For this exercise, you will use a pre-existing projection/coordinate system called British National Grid. This is a transverse mercator projection which is specifically designed to minimise distortion of features in the British Isles and it uses the OSGB 1936 datum. To set your data frame to use the British National Grid projection/coordinate system, work through the following flow diagram:

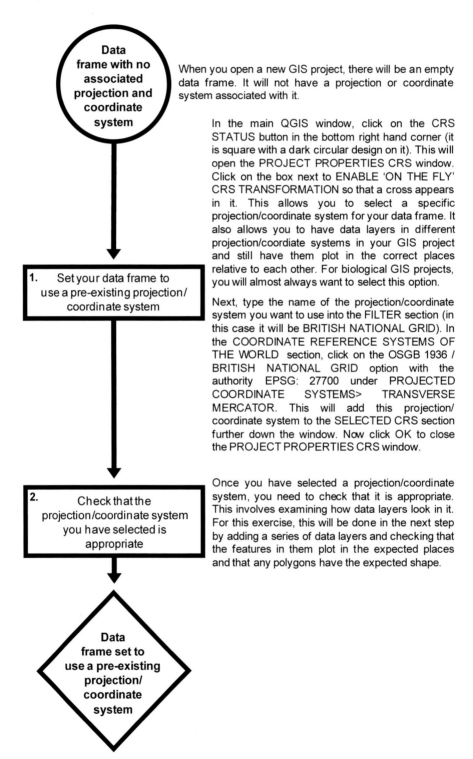

Data frame with no associated projection and coordinate system

When you open a new GIS project, there will be an empty data frame. It will not have a projection or coordinate system associated with it.

In the main QGIS window, click on the CRS STATUS button in the bottom right hand corner (it is square with a dark circular design on it). This will open the PROJECT PROPERTIES CRS window. Click on the box next to ENABLE 'ON THE FLY' CRS TRANSFORMATION so that a cross appears in it. This allows you to select a specific projection/coordinate system for your data frame. It also allows you to have data layers in different projection/coordiate systems in your GIS project and still have them plot in the correct places relative to each other. For biological GIS projects, you will almost always want to select this option.

1. Set your data frame to use a pre-existing projection/ coordinate system

Next, type the name of the projection/coordinate system you want to use into the FILTER section (in this case it will be BRITISH NATIONAL GRID). In the COORDINATE REFERENCE SYSTEMS OF THE WORLD section, click on the OSGB 1936 / BRITISH NATIONAL GRID option with the authority EPSG: 27700 under PROJECTED COORDINATE SYSTEMS> TRANSVERSE MERCATOR. This will add this projection/ coordinate system to the SELECTED CRS section further down the window. Now click OK to close the PROJECT PROPERTIES CRS window.

2. Check that the projection/coordinate system you have selected is appropriate

Once you have selected a projection/coordinate system, you need to check that it is appropriate. This involves examining how data layers look in it. For this exercise, this will be done in the next step by adding a series of data layers and checking that the features in them plot in the expected places and that any polygons have the expected shape.

Data frame set to use a pre-existing projection/ coordinate system

To check that you have done this step correctly, click on PROJECT on the main menu bar and select PROJECT PROPERTIES. In the PROJECT PROPERTIES window which will open, click on the CRS tab on the left hand side and make sure that the contents of the SELECTED CRS section has the following text in it:

OSGB 1936/ British National Grid

Underneath this, it should say:

+proj=tmerc +lat_0=49 +lon_0=-2 +k=0.9996012717 +x_0=400000 +_0=- 100000 +ellps=airy +towgs84=446.448,-125.157,542.06,0.15,0.247,0.842,- 20.489 +units=m +no_defs

This is known as the Proj.4 string, which is used to define the characteristics of the projection/coordinate system.

If it does, click OK to close the PROJECT PROPERTIES window. If it does not, you will need to repeat this step until you have assigned the correct projection/coordinate system to your data frame. Once you have successfully completed this step, click on the PROJECT menu on the main menu bar and select SAVE to save the changes you have made to your GIS project.

STEP 2: ADD ANY EXISTING DATA LAYERS YOU WISH TO DISPLAY ON YOUR MAP TO YOUR GIS PROJECT:

Once you have set the projection/coordinate system of your data frame, you are ready to add some existing data layers to it. For this exercise, you will use a number of existing data layers which provide information about a variety of different features of the native oak woodland study area. This will include the area of the woodland itself (SCENE_OAK_WOODLAND), information about the position of the shoreline of a large lake (LOCH_LOMOND_ SHORELINE) and information about nearby roads (SCENE_ROADS). These existing data layers are all in the British National Grid projection which will be used for this exercise.

To add these data layers to your GIS project, work through the flow diagram which starts on the next page.

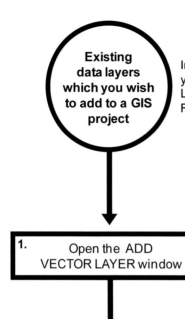

In this exercise, the data layers which you wish to add to your GIS project are SCENE_OAK_WOODLAND.SHP, LOCH_LOMOND_SHORELINE.SHP and SCENE_ROADS.SHP.

1. Open the ADD VECTOR LAYER window

On the main menu bar, click on LAYER and select ADD LAYER> ADD VECTOR LAYER. In the ADD VECTOR LAYER window, browse to the location of your data layer (C:\GIS_FOR_BIOLOGISTS) and then click on the section in the bottom right hand corner of the OPEN window (where it currently says ALL FILES (*).(*)) and select ESRI SHAPEFILES (*.shp, .SHP). Now, select the data layer called SCENE_OAK_WOODLAND.SHP. Next, click OPEN in the browse window and then OPEN in the ADD VECTOR LAYER window.

2. Check the projection/coordinate system for the newly added data layer

Whenever you add a data layer to a GIS project, you should always check that it has a projection/coordinate system assigned to it, and look at what this projection/coordinate system is. This is so that you know whether you will need to assign a projection/coordinate system to it, or transform it into a different projection/coordinate system before you can use it in your GIS project. To check the projection/coordinate system of your newly added data layer, right click on its name in the TABLE OF CONTENTS window and select PROPERTIES. In the LAYER PROPERTIES window which opens, click on the GENERAL tab on the left hand side and check that there is a projection/coordinate system listed in the COORDINATE REFERENCE SYSTEM window. For the SCENE_OAK_WOODLAND data layer this should say :

EPSG:27700 – OSGB 1936 / British National Grid

This tells you that this data layer is in the British National Grid projection based on the OSGB 1936 datum. Now click OK to close the LAYER PROPERTIES window.

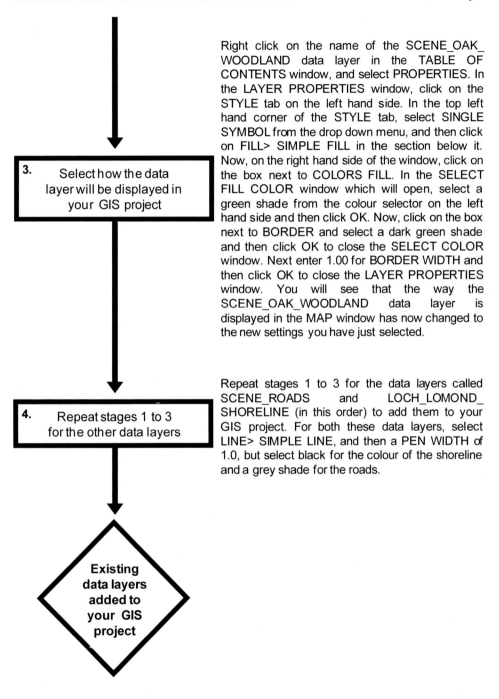

3. Select how the data layer will be displayed in your GIS project

Right click on the name of the SCENE_OAK_ WOODLAND data layer in the TABLE OF CONTENTS window, and select PROPERTIES. In the LAYER PROPERTIES window, click on the STYLE tab on the left hand side. In the top left hand corner of the STYLE tab, select SINGLE SYMBOL from the drop down menu, and then click on FILL> SIMPLE FILL in the section below it. Now, on the right hand side of the window, click on the box next to COLORS FILL. In the SELECT FILL COLOR window which will open, select a green shade from the colour selector on the left hand side and then click OK. Now, click on the box next to BORDER and select a dark green shade and then click OK to close the SELECT COLOR window. Next enter 1.00 for BORDER WIDTH and then click OK to close the LAYER PROPERTIES window. You will see that the way the SCENE_OAK_WOODLAND data layer is displayed in the MAP window has now changed to the new settings you have just selected.

4. Repeat stages 1 to 3 for the other data layers

Repeat stages 1 to 3 for the data layers called SCENE_ROADS and LOCH_LOMOND_ SHORELINE (in this order) to add them to your GIS project. For both these data layers, select LINE> SIMPLE LINE, and then a PEN WIDTH of 1.0, but select black for the colour of the shoreline and a grey shade for the roads.

Existing data layers added to your GIS project

At the end of this step, your TABLE OF CONTENTS window should look like the image at the top of the next page.

137

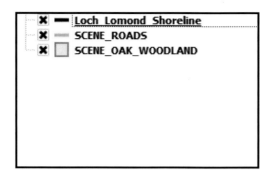

While the contents of your MAP window should look like this:

If it does not look like this, try right-clicking on the name of the SCENE_OAK_WOODLAND data layer in the TABLE OF CONTENTS window and selecting ZOOM TO LAYER. If it still does not look right, remove all the data layers from your GIS project by right-clicking on their names in the TABLE OF CONTENTS window and selecting REMOVE. Next, go back to step one and ensure that you have set the projection/coordinate system of your data frame correctly, and then repeat step two. Once you have successfully completed this step, click on the PROJECT menu on the main menu bar and select SAVE to save the changes you have made to your GIS project.

STEP 3: CREATE A NEW POINT DATA LAYER FROM A LIST OF COORDINATES IN A SPREADSHEET OR DATABASE:

Now that you have added the required existing data layers to your GIS project, you can start creating new ones to add further information to your map. The first new data layer you will create will be a point data layer of the nest box locations used to study the breeding success of hole-nesting birds in the oak woodland from a list of coordinates held in a spreadsheet. This will be done in two parts. Firstly, you will plot the locations of all the nest boxes in the spreadsheet and make them into a point data layer. Once you have done this, you will select only those nest boxes which are within the oak woodland study area and make a new data layer with just these locations in it. Therefore, in this step, you will learn two different ways of making a new data layer from existing data sets. To create your new data layer from a list of coordinates held in a spreadsheet or database, you would first need to save the information as a tab delimited text file and then work through the following flow diagram:

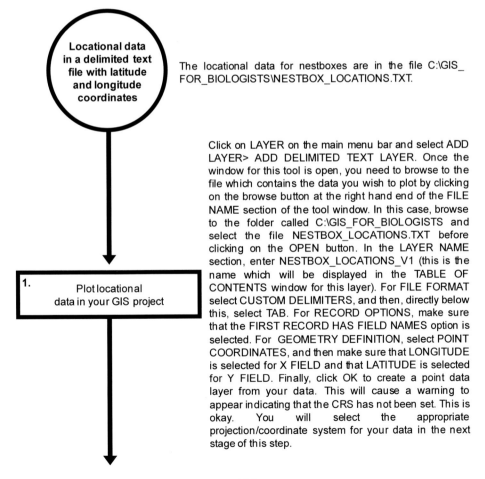

Locational data in a delimited text file with latitude and longitude coordinates

The locational data for nestboxes are in the file C:\GIS_FOR_BIOLOGISTS\NESTBOX_LOCATIONS.TXT.

1. Plot locational data in your GIS project

Click on LAYER on the main menu bar and select ADD LAYER> ADD DELIMITED TEXT LAYER. Once the window for this tool is open, you need to browse to the file which contains the data you wish to plot by clicking on the browse button at the right hand end of the FILE NAME section of the tool window. In this case, browse to the folder called C:\GIS_FOR_BIOLOGISTS and select the file NESTBOX_LOCATIONS.TXT before clicking on the OPEN button. In the LAYER NAME section, enter NESTBOX_LOCATIONS_V1 (this is the name which will be displayed in the TABLE OF CONTENTS window for this layer). For FILE FORMAT select CUSTOM DELIMITERS, and then, directly below this, select TAB. For RECORD OPTIONS, make sure that the FIRST RECORD HAS FIELD NAMES option is selected. For GEOMETRY DEFINITION, select POINT COORDINATES, and then make sure that LONGITUDE is selected for X FIELD and that LATITUDE is selected for Y FIELD. Finally, click OK to create a point data layer from your data. This will cause a warning to appear indicating that the CRS has not been set. This is okay. You will select the appropriate projection/coordinate system for your data in the next stage of this step.

2. Assign the correct projection/coordinate system to the data layer

Right click on the data layer called NESTBOX_ LOCATIONS_V1 in the TABLE OF CONTENTS and select PROPERTIES. In the LAYER PROPERTIES window, click on the GENERAL tab and then click on the SELECT CRS button at the right hand end of the COORDINATE REFERENCE SYSTEM section (it has a picture of a globe and small yellow square on it). In the COORDINATE REFERENCE SYSTEM SELECTOR window which will open, type WGS 84 into the FILTER section (making sure that you leave a space between the letters WGS and the number 84). Next, in the COORDINATE REFERENCE SYSTEMS OF THE WORLD section, select WGS 84 under GEOGRAPHIC COORDINATE SYSTEM. **NOTE:** If your coordinates were in a different projection/ coordinate system, you would set this different system at this stage. In the SELECTED CRS section, you should now see WGS 84, and in the window below it the following proj.4 string:

```
+proj=longlat +datum=WGS84 +no_defs
```

Finally, click OK to set this as the projection/coordinate system for the point data layer you have just created, and then click OK to close the LAYER PROPERTIES window.

3. Transform your data layer into the projection/ coordinate system being used for your GIS project and save it as a shapefile

In order to make a permanent version of your data layer, you need to convert it into a shapefile. This can be done using the SAVE AS tool. To access this tool, right click on the name of the data layer you just created in the TABLE OF CONTENTS window, and select SAVE AS. This opens the SAVE VECTOR LAYER AS. In this window, for FORMAT select ESRI SHAPEFILE. Next, enter the following address and file name into the SAVE AS section of the window: C:/GIS_FOR_BIOLOGISTS/NESTBOX_LOCATIONS_ V2.SHP. **NOTE:** This address uses slashes (/) rather than the usual backslashes (\). In the CRS section, make sure that the option starting with PROJECT CRS is selected. Now, select ADD SAVED FILE TO MAP and then click on the OK button.

Finally, right click on the name of the data layer you created in stages 1 and 2 (NESTBOX_LOCATIONS_ V1) in the TABLE OF CONTENTS window and select REMOVE from the menu which appears to remove it from your GIS project. This is because you no longer need it now that you have transformed your data layer into the projection/coordinate system you are using for your project and have converted it into a shapefile.

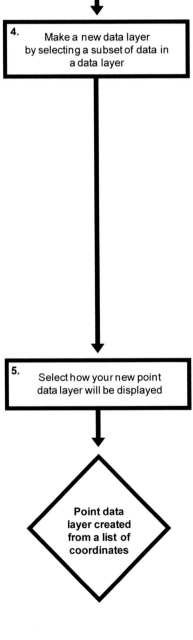

4. Make a new data layer by selecting a subset of data in a data layer

5. Select how your new point data layer will be displayed

Point data layer created from a list of coordinates

If you right click on the name of the data layer you have just created (NESTBOX_LOCATIONS_V2) in the TABLE OF CONTENTS window and select ZOOM TO LAYER, you will see that there are two groups of nest boxes in it, one set in the SCENE_OAK_WOODLAND polygon and one further south. You now want to select those nest boxes in the SCENE oak woodland and make a new data layer with just these locations in it. To do this, click on VECTOR on the main menu bar and select SPATIAL QUERY> SPATIAL QUERY (**NOTE**: if this option is not available, check that the SPATIAL the SPATIAL QUERY plugin has been activated – see note on page 133 for more information). This will open the SPATIAL QUERY window. For SELECT SOURCE FEATURES FROM, select NESTBOX_ LOCATIONS_V2 from the drop down menu. For REFERENCE FEATURES OF, select SCENE_OAK_WOODLAND, and then for WHERE THE FEATURE, select WITHIN. Now click APPLY to run the spatial query and then CLOSE to close the SPATIAL QUERY tool window.

To make a new data layer containing only the selected locations, right click on the data layer called NESTBOX_LOCATIONS_V2 in the TABLE OF CONTENTS window and select SAVE AS. This opens the SAVE VECTOR LAYER AS window. In this window, for FORMAT select ESRI SHAPEFILE. Next, enter the following address and file name into the SAVE AS section of the window: C:/GIS_FOR_BIOLOGISTS/ SCENE_NESTBOX_LOCATIONS.SHP. **NOTE:** This address uses slashes (/) rather than the usual backslashes (\). In the CRS section, make sure that the option starting with LAYER CRS is selected. Now, select SAVE ONLY SELECTED FEATURES and then select ADD SAVED FILE TO MAP. Finally, to save the selected records as a new data layer, click on the OK button.

Right click on the name of the SCENE_NESTBOX_ LOCATIONS data layer in the TABLE OF CONTENTS window, and select PROPERTIES. In the LAYER PROPERTIES window, click on the STYLE tab. In the top left hand corner of the STYLE tab, select SINGLE SYMBOL from the drop down menu. Next, in the section on the right hand side of the STYLE tab, select CIRCLE. Once this has been selected, enter 2 for SIZE before clicking on the box next to COLOR and selecting a red shade in the SELECT COLOR window which will open. Finally, click OK to close the SELECT COLOR window and then OK to close the LAYER PROPERTIES window.

At the end of this step, your TABLE OF CONTENTS window should look like this:

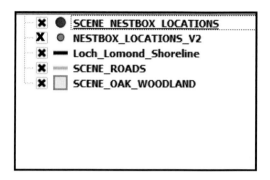

Click on the box next to the name of the data layer SCENE_NESTBOX_ LOCATIONS in the TABLE OF CONTENTS window so that it is no longer displayed in the MAP window. Next, click on VIEW on the main menu bar and select SELECT> DESELECT FEATURES FROM ALL LAYERS. The contents of your MAP window should now look like this:

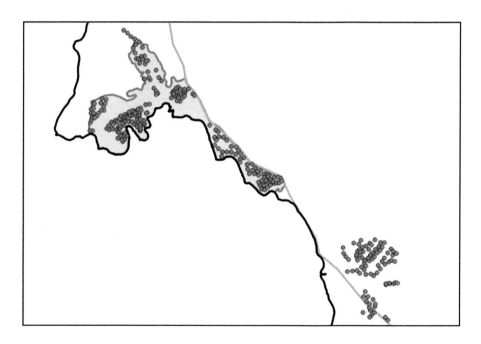

Now check the box next to SCENE_NESTBOX_LOCATIONS in the TABLE OF CONTENTS window again (so that it is displayed in the MAP window once more) and uncheck the box next to NESTBOX_ LOCATIONS_V2 (so that it is no longer displayed in the MAP window). You will now see that it contains only those nest box locations in the SCENE oak woodland. Right click on SCENE_OAK_WOODLAND in the TABLE OF CONTENTS window and select ZOOM TO LAYER. The contents of your MAP window should now look like this:

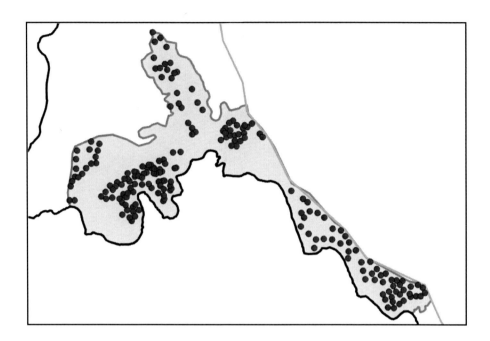

Once you have successfully completed this step, click on the PROJECT menu on the main menu bar and select SAVE to save the changes you have made to your GIS project.

STEP 4: CREATE A NEW DATA LAYER WITHIN YOUR GIS SOFTWARE:

In step three, you learned how to make new data layers in two different ways. In this step, you will learn how to make a new data layer in a third way. This is directly in the MAP window of your GIS software. You can also make polygon and line data layers in this way, but the data layer you will create in this step will be a point data layer. This will have a single point marking the location of the University of Glasgow's SCENE field station buildings and will be called SCENE_BUILDINGS. This data layer can be made by working through the flow diagram which starts on the next page.

NOTE: This step requires a plug-in to be loaded into QGIS from the internet. This means that you will need to have an internet connection. In addition, if you are connected to an organisational network, you may need to change your proxy setting by clicking on SETTINGS on the main menu bar and selecting NETWORK. To find out what proxy settings you should use, speak to you network's IT support.

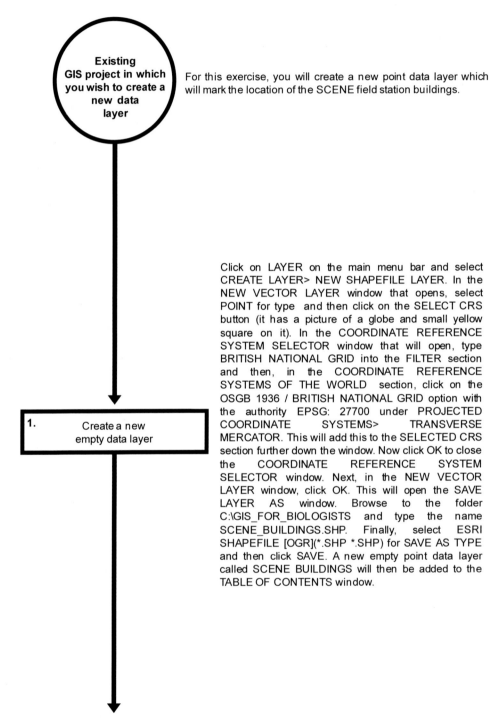

Existing GIS project in which you wish to create a new data layer

For this exercise, you will create a new point data layer which will mark the location of the SCENE field station buildings.

1. Create a new empty data layer

Click on LAYER on the main menu bar and select CREATE LAYER> NEW SHAPEFILE LAYER. In the NEW VECTOR LAYER window that opens, select POINT for type and then click on the SELECT CRS button (it has a picture of a globe and small yellow square on it). In the COORDINATE REFERENCE SYSTEM SELECTOR window that will open, type BRITISH NATIONAL GRID into the FILTER section and then, in the COORDINATE REFERENCE SYSTEMS OF THE WORLD section, click on the OSGB 1936 / BRITISH NATIONAL GRID option with the authority EPSG: 27700 under PROJECTED COORDINATE SYSTEMS> TRANSVERSE MERCATOR. This will add this to the SELECTED CRS section further down the window. Now click OK to close the COORDINATE REFERENCE SYSTEM SELECTOR window. Next, in the NEW VECTOR LAYER window, click OK. This will open the SAVE LAYER AS window. Browse to the folder C:\GIS_FOR_BIOLOGISTS and type the name SCENE_BUILDINGS.SHP. Finally, select ESRI SHAPEFILE [OGR](*.SHP *.SHP) for SAVE AS TYPE and then click SAVE. A new empty point data layer called SCENE BUILDINGS will then be added to the TABLE OF CONTENTS window.

In order to allow you to enter the exact coordinates where you wish a point (or vertex of a line or polygon) to be located, you will need to load the NUMERICAL VERTEX EDITOR plugin. To load this plugin, first click on PROJECT on the main menu bar an select SAVE. Next, click on PLUGINS on the main menu bar and select MANAGE AND INSTALL PLUGINS. In the PLUGINS window, click on ALL and then select NUMERICAL VERTEX EDIT and click INSTALL PLUGIN. Once the plugin has been installed, click CLOSE to close the PLUGINS window. You now need to re-open your GIS project in order to activate the plugin. To do this, click on PROJECT on the main menu bar and select OPEN RECENT> C:/GIS_FOR_ BIOLOGISTS/EXERCISE_TWO.QGIS.

2. Start adding features to your new data layer

Next, right click on the name of your new data layer (i.e. SCENE_BUILDINGS) in the TABLE OF CONTENTS window and select TOGGLE EDITING. Next, click on EDIT on the main menu bar and select ADD FEATURE. Now click anywhere in the MAP window. This will add a new point to the data layer you are editing, and will open the FEATURE ATTRIBUTE WINDOW (titled SCENE_BUILDI...). In this window, enter a value of 1 and click OK. Now click on EDIT and select NUMERICAL VERTEX EDIT. In the MAP window, move the cursor directly over the point you have just created and click on it. This will open the MOVE VERTEX FEATURE window where you can enter the specific coordinates you want to position your point at (these will be in the coordinate system of the projection/coordinate system which your GIS project is in). For the location of the SCENE buildings point, enter 237665, 695980, and then click OK. This will move the point to these specific coordinates.

3. Add information to the attribute table of your new data layer

To add information to the attribute table of your new data layer, right click on the data layer's name in the TABLE OF CONTENTS window and select OPEN ATTRIBUTE TABLE. In the ATTRIBUTE TABLE window, click on the NEW COLUMN button (it is the one second from the right hand end of the row of buttons at the top of the ATTRIBUTE TABLE window). In the ADD COLUMN window which will open, enter LOCATION in the NAME section and select TEXT (STRING) for type before entering 50 for WIDTH, and then clicking OK. Now, in the ATTRIBUTE TABLE window, double click on the first cell in the newly added LOCATION field and type SCENE BUILDINGS before pressing the ENTER key on your keyboard. Finally, click on the SAVE EDITS button (it has a picture of an old-fashioned diskette on it) and then click the TOGGLE EDIT button (it has a yellow pencil on it) to close the editing session.

145

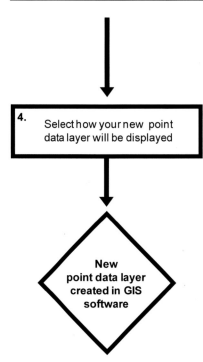

Return to the MAP window before right-clicking on the name of the SCENE_BUILDINGS data layer in the TABLE OF CONTENTS window, and selecting PROPERTIES. In the LAYER PROPERTIES window, click on the STYLE tab. In the top left hand corner of the STYLE tab, select SINGLE SYMBOL from the drop down menu. Next, in the section on the right hand side of the STYLE tab, select SQUARE. Once this has been selected, enter 2 for SIZE before clicking on the box next to COLOR and selecting the colour black in the SELECT COLOR window which will open. Finally, click OK to close the SELECT COLOR window and then OK to close the LAYER PROPERTIES window.

At the end of this step, the contents of your ATTRIBUTE TABLE window should look like this:

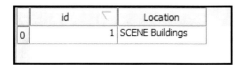

Now close the ATTRIBUTE TABLE window. Your TABLE OF CONTENTS window should look like this:

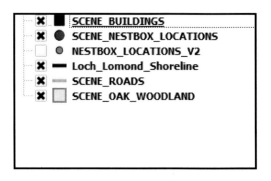

Finally, click on VIEW on the main menu bar and select SELECT> DESELECT FEATURES FROM ALL LAYERS. The contents of your MAP window should now look like this:

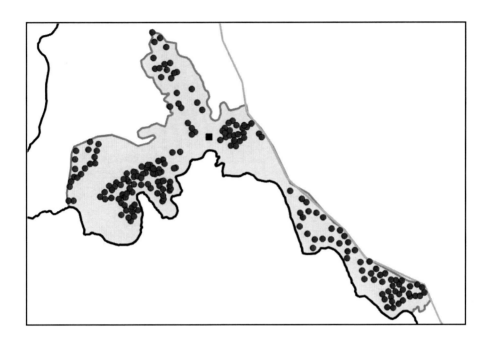

Once you have successfully completed this step, click on the PROJECT menu on the main menu bar and select SAVE to save the changes you have made to your GIS project.

STEP 5: CREATE A NEW POLYGON DATA LAYER BY TRACING A FEATURE FROM A GOOGLE EARTH IMAGE:

Google Earth images provide high quality aerial images covering much of the world, and this means that they are a great source of information which biologists can use for many different purposes. In this step, you will learn how to create a polygon data layer by tracing a feature in a Google Earth image and then importing it into your GIS project from the Google Earth data layer format (which has the extension .kml or .kmz). The feature you will trace is a small lake, called the Dubh Loch, which borders on the oak woodland study area. **NOTE:** To be able to do this step, you will need to have a copy of the Google Earth user interface installed on your computer, and you will need to have access to the internet. If you do not have either of these, you can skip this step and move on to step 6. The polygon data layer of the Dubh Loch can be made by working through the flow diagram which starts at the top of the next page.

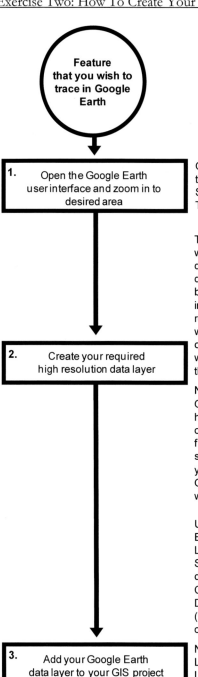

Feature that you wish to trace in Google Earth

1. Open the Google Earth user interface and zoom in to desired area

2. Create your required high resolution data layer

3. Add your Google Earth data layer to your GIS project

Open the Google Earth user interface. Once it opens, type DUBH LOCHAN, GLASGOW, G63, UK into the SEARCH window and then click the SEARCH button. This will allow you to locate the Dubh Loch.

To create a high resolution data layer, first decide whether you want to create a point data layer, a line data layer or a polygon data layer. Once you have decided this, select the appropriate tool from the buttons on the main menu bar of the Google Earth user interface. For this exercise, you will make a high resolution polygon data layer of the Dubh Loch, so you will need to select the ADD POLYGON tool by clicking on ADD and selecting POLYGON. This will open a window called GOOGLE EARTH – NEW POLYGON. In this window, type DUBH LOCH beside NAME.

Now move this window so that you can see all of the Google Earth MAP window. Starting at the bottom left hand corner, trace out the outline of the Dubh Loch by clicking at regular intervals along its shoreline. If you find that you have added a point in the wrong place, simply press the DELETE key on your keyboard. When you are happy with the polygon you have drawn, click OK to close the GOOGLE EARTH – NEW POLYGON window.

Under PLACES on the left hand side of the Google Earth user interface, right click on the name DUBH LOCH and select SAVE PLACE AS. This will open the SAVE FILE window. **NOTE:** You may have to scroll down to find this data layer. Browse to the folder C:\GIS_FOR_BIOLOGISTS and then type in the name DUBH_LOCH. Next, under SAVE AS TYPE select KML (*.KML) and then click the SAVE button. You can now close the Google Earth user interface.

Next, return to QGIS. On the main menu bar, click on LAYER and select ADD LAYER> ADD VECTOR LAYER. In the ADD VECTOR LAYER window, browse to the location of your data layer (C:\GIS_FOR_ BIOLOGISTS) before clicking on the section in the bottom right hand corner of the OPEN window (where it currently says ESRI SHAPEFILE (*.SHP).(*.SHP)) and selecting KEYHOLE MARKUP LANGUAGE [KML] (*.KML *.KMZ) from the drop down menu that will appear. Now, select the data layer called DUBH_LOCH.KML. Next, click OPEN in the browse window and then OPEN in the ADD VECTOR LAYER window.

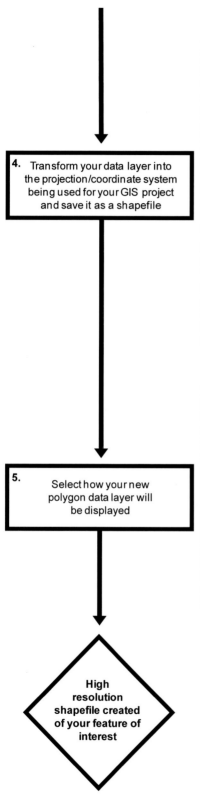

4. Transform your data layer into the projection/coordinate system being used for your GIS project and save it as a shapefile

5. Select how your new polygon data layer will be displayed

High resolution shapefile created of your feature of interest

In order to make a permanent version of your data layer, you need to convert it into a shapefile. This can be done using the SAVE AS tool. To access this tool, right click on the name of the data layer you have just created in the TABLE OF CONTENTS window, and select SAVE AS. This opens the SAVE VECTOR LAYER AS window. In this window, for FORMAT select ESRI SHAPEFILE. Next, enter the following address and file name into the SAVE AS section of the window: C:/GIS_FOR_BIOLOGISTS/DUBH_LOCH.SHP. **NOTE:** This address uses slashes (/) rather than the usual backslashes (\). In the CRS section, make sure that the option starting with PROJECT CRS is selected. Now, select ADD SAVED FILE TO MAP and then click on the OK button.

Finally, right click on the name of the data layer you created in stages 1 and 2 (DUBH_LOCH.KML) in the TABLE OF CONTENTS window and select REMOVE from the menu which appears. This will remove this data layer from your GIS project. This is okay as you no longer need it now you have transformed your data layer into a shapfile based on the projection/coordinate system you are using for your project.

Right click on the name of the DUBH_LOCH data layer in the TABLE OF CONTENTS window, and select PROPERTIES. In the LAYER PROPERTIES window, click on the STYLE tab on the left hand side. In the top left hand corner of the STYLE tab, select SINGLE SYMBOL and then click on FILL> SIMPLE FILL in the section below it. Now, on the right hand side of the window, click on the box next to COLORS FILL. In the SELECT COLOR window which will open, select a light blue shade and click OK. Now click on the box next to BORDER and select a dark blue shade, then click OK to close the SELECT COLOR window. Next enter 0.5 for BORDER WIDTH and then OK to close the LAYER PROPERTIES window. You will see that the way the DUBH_LOCH data layer is displayed in the MAP window has now changed to the new settings you have just selected.

At the end of this step, in the TABLE OF CONTENTS window, click on the DUBH_LOCH data layer and drag it down until it is between the SCENE_ROADS data layer and SCENE_OAK_WOODLAND data layer. Your TABLE OF CONTENTS window should now look like this:

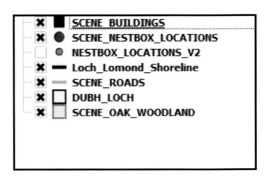

While the contents of your MAP window should look like this:

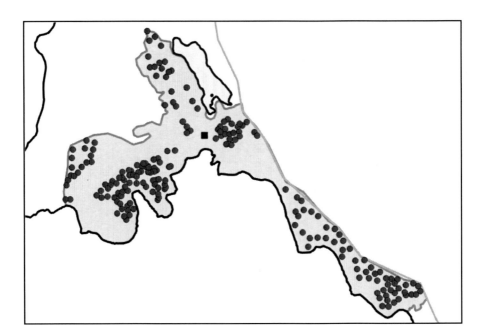

Once you have successfully completed this step, click on the PROJECT menu on the main menu bar and select SAVE to save the changes you have made to your GIS project.

STEP 6: CREATE A NEW LINE DATA LAYER BY IMPORTING A TRACK RECORDED ON A GPS RECEIVER:

The final way you will create a new data layer in this exercise is from data recorded on a GPS receiver. GPS receivers can record two types of data: waypoints, which mark specific locations, and tracks, which are lines marking specific routes. You can use such tracks to record features which you might want to add to a GIS project simply by setting up your GPS to record a track and then walking along the feature at a steady pace. Once you reach the end of the feature, you can save the section of track you have just recorded and then load it into a GIS project. In this step, you will do this for a GPS track file (which, like all GPS compatible files, has the extension .gpx) that was recorded by walking along a forestry track which runs through the middle of the oak woodland study area. This will allow you to add this feature to the map you are building of the oak woodland study area. The line data layer of the forestry track can be created by working through the flow diagram at the bottom of this page. To find out more about how to set up a GPS to record a track, visit *www.GISinEcology.com/GFB.htm#video9*.

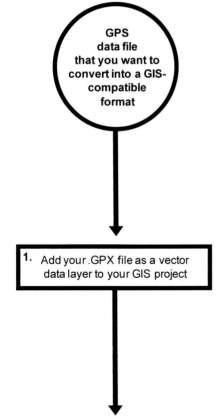

On the main menu bar, click on LAYER and select ADD LAYER> ADD VECTOR LAYER. In the ADD VECTOR LAYER window, browse to the location of your data layer (C:\GIS_FOR_BIOLOGISTS). Now, click on the section in the bottom right hand corner of the OPEN window (where it currently says KEYHOLE MARKUP LANGUAGE [KML] (*.KML *.KMZ) and select GPS EXCHANGE FORMAT [GPX] (*.GPX*.GPX). Next, select the data layer called FORESTRY_TRACK.GPX. Click OPEN in the BROWSE window and then OPEN in the ADD VECTOR LAYER window. This will open the SELECT VECTOR LAYERS TO ADD window. In this window, select the line which starts 2 TRACKS and click OK.

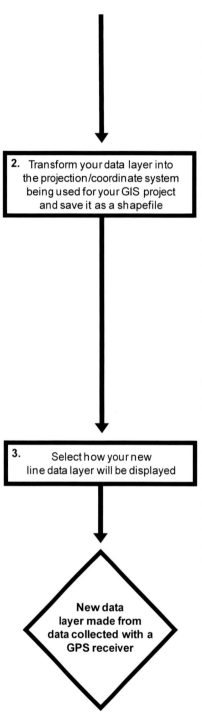

2. Transform your data layer into the projection/coordinate system being used for your GIS project and save it as a shapefile

3. Select how your new line data layer will be displayed

New data layer made from data collected with a GPS receiver

In order to make a permanent version of your data layer, you need to convert it into a shapefile. This can be done using the SAVE AS tool. To access this tool, right click on the name of the data layer you just created in the TABLE OF CONTENTS window, and select SAVE AS. In the SAVE VECTOR LAYER AS window which will open, select ESRI SHAPEFILE, for FORMAT. Next, enter C:/GIS_FOR_BIOLOGISTS/ FORESTRY_TRACK.SHP into the SAVE AS section of the window. **NOTE**: This address uses slashes (/) rather than the usual backslashes (\). In the CRS section, make sure that you select the option that starts with PROJECT CRS. Now, select ADD SAVED FILE TO MAP and then click on the OK button.

Finally, right click on the name of the data layer you created in stage 1 (TRACKS ANY) in the TABLE OF CONTENTS window and select REMOVE from the menu which appears. This will remove this data layer from your GIS project. This is okay as you no longer need it now you have transformed your data layer into a shapefile based on the projection/coordinate system you are using for your project.

Right click on the name of the FORESTRY_TRACK data layer in the TABLE OF CONTENTS window, and select PROPERTIES. In the LAYER PROPERTIES window, click on the STYLE tab on the left hand side. In the top left hand corner of the STYLE tab, select SINGLE SYMBOL from the drop down menu, and then click on LINE> SIMPLE LINE in the section below it. Now, on the right hand side of the window, click on the box next to COLOR. In the SELECT COLOR window which will open, select a brown shade and click OK. Now enter 0.5 for PEN WIDTH and then OK to close the LAYER PROPERTIES window. You will see that the way the FORESTRY_TRACK data layer is displayed in the MAP window has now changed to the new settings you have just selected.

At the end of this step, in the TABLE OF CONTENTS window, click on the FORESTRY_TRACK data layer and drag it down until it is between the

NESTBOX_LOCATIONS data layer and LOCH_LOMOND_SHORLINE data layer. Your TABLE OF CONTENTS window should now look like this:

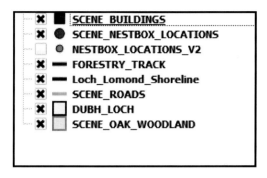

While the contents of your MAP window should look like this:

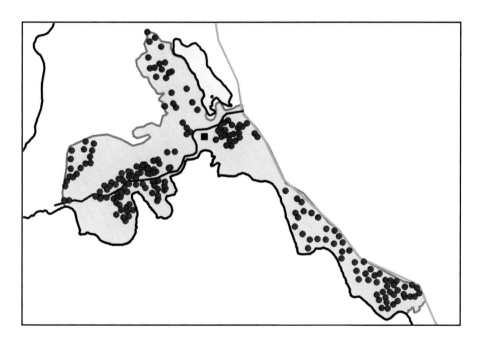

Once you have successfully completed this step, click on PROJECT on the main menu bar and select SAVE to save the changes you have made to your GIS project.

This is the end of exercise two. Now that you have completed it, you know all the main ways that you can make new feature data layers in a GIS project. To recap, these are: 1. Creating a point data layer from a list of coordinates stored in a spreadsheet or database; 2. By selecting a subset of data in an existing data layer; 3. Creating a new data layer directly in your GIS software; 4. Creating a data layer by tracing a feature in Google Earth image; 5. Creating a data layer using a track or waypoint file recorded on a GPS receiver.

--- Chapter Fourteen ---

Exercise Three: How To Work With Raster Data Layers

In exercises one and two, you worked with feature data layers. In exercise three, you will start working with raster data layers for the first time. Raster data layers are gridded data sets and they are commonly used to create continuous surfaces of environmental information, such as elevation or water depth, or to provide information about the distribution of species, such as maps of species richness (see exercise six in chapter seventeen). As outlined below, raster data layers differ from feature data layers in a number of key ways.

Feature (Vector) Data Layers	Raster (Gridded) Data Layers
• Consist of individual points, lines or polygons (features).	• Consist of a grid with cells of a specific size.
• Have an attribute table which is used to store information about each feature.	• Information about the values for each grid cell is stored in an array.
• Can have information about multiple variables for each point in space.	• Can usually only have information about one variable for each point in space.
• Individual features can overlap each other.	• Individual cells cannot overlap.
• Individual features can be edited.	• Individual features cannot be edited.

In this exercise, you will create raster data layers for a number of different environmental variables for a study area centred on Mount Mabu in northern Mozambique. This will give you experience in creating raster data layers from feature data layers, and in deriving additional data layers from existing raster data layers. While this exercise concentrates on creating raster data layers of environmental variables (one of the main uses for raster data layers in biology), you will find other examples of how raster data layers can be created and used in exercises five (chapter sixteen) and six (chapter seventeen).

Mount Mabu is a remote and isolated mountain peak which rises to a height of approximately 1,700 metres (~5,600 feet) above sea level and which is clothed in rainforest. It towers above the surrounding savannah and, as a result, is isolated from other areas of similar habitat. This means that it is home to a high diversity of species which are not found anywhere else. It is also unusual in the fact that to date (2015), it remains relatively unaffected by human activities, and so its rainforest ecosystem remains relatively intact. The existence of the rainforest on Mount Mabu has only recently become known to scientists (although well-known to locals prior to this, it was 'discovered' by the outside world in 2005 when scientists working at Kew Royal Botanic Gardens spotted it in a Google Earth image). It remains very poorly studied and raster data layers of this area could be used to help identify potentially interesting habitats where specific types of species might occur, and so which are worthy of more detailed investigation. Thus, in this exercise, you will create raster data layers of elevation, the steepness of the slope, how much direct sunlight each location gets and the distance from each point in the forest to the nearest edge, with the aim of identifying areas where new species with specific habitat preferences are likely to occur.

In addition, while Mount Mabu has not, to date, been affected to any great extent by human activities, this does not mean that it is not vulnerable to over-exploitation. In particular, as with any rainforest area, there is the potential for deforestation. This means that it is important to monitor the extent of the rainforest which covers the top of this mountain so that any evidence of deforestation can be identified as soon as possible to allow for its effects to be mitigated. To do this, as the final step in this exercise, you will create a raster data layer of the current extent of the rainforest which could be used as a baseline for future environmental monitoring.

Thus, the raster data layers you will create in this exercise are:

1. **Elevation:** This raster data layer will provide information on the altitude of each 100m by 100m grid cell covering the whole of Mount Mabu. It will be generated from a line data layer of elevation contours and it will form a digital elevation model (or DEM for short).

2. **Slope:** This raster data layer will provide information on how the steepness of the slope varies across Mount Mabu and can be used to identify areas of steep and gentle slopes. It will be generated from the elevation raster data layer.

3. **Hillshade:** This raster data layer will provide information on how much sunlight each grid cell will receive at a specific time of day and time of year. This will affect whether shade or sun-loving species are likely to be found

within a specific grid cell. It will be generated from the elevation raster data layer.

4. **Distance From The Forest Edge:** This raster data layer will provide information on how far each grid cell in the rainforest is from the forest edge. This will affect where species which prefer to live near or avoid edge habitats are likely to occur. It will be generated from a line data layer representing the outer edge of the rainforest area and a polygon representing the total area covered by the forest.

5. **Area of Forest:** This raster data layer will provide a measure of the extent of the rainforest on Mount Mabu in 2013. It can be used to monitor the rainforest over time and identify any areas where deforestation might be taking place. It will be generated from a polygon representing the rainforest which was made from a Google Earth image (see exercise two in chapter thirteen for details of how this type of data layer can be created).

When creating raster data layers, there are three crucial elements which must be considered. These are the projection/coordinate system that it will be made in, its extent, and the cell size or resolution. This means it is important that you use an appropriate choice for each of these elements. The appropriate choice will depend on the location in the world your raster data layer will cover, the size of the area it will cover, the resolution of the data layers you will make it from and exactly what you are wanting to do with your raster data layers. If you use the wrong settings when you create your raster data layers, then anything which you later do with them will be incorrect. This means that it is important to get these settings right from the start.

For this exercise, the raster data layers will be created in the Universal Transverse Mercator (UTM) Zone 37S projection based on the WGS 1984 datum. They will use an extent which has coordinates for its top left corner of 207000 E, 8214000 N and 227000 E, 8191000 N for its bottom right corner (**NOTE:** These coordinates are in easting and northing for the UTM Zone 37S projection/coordinate system and not latitude/longitude). The raster data layers will have a resolution of 100m (meaning each grid cell represents a square that is 100m by 100m). These values have been selected on the basis of the size of the area which will be covered, its location in the world and the resolution of the original feature data sets from which the raster data layers will be created (see below). The instructions for creating the above raster data layers for Mount Mabu in ArcGIS 10.3 can be found on page 159, while those for creating the raster data layers in QGIS 2.8.3 can be found on page 180.

If you have not done so already, before you start this exercise, you will first need to create a new folder on your C: drive called GIS_FOR_BIOLOGISTS.

To do this on a computer with a Windows operating system, open Windows Explorer and navigate to your C:\ drive (this may be called Windows C:). To create a new folder on this drive, right click on the window displaying the contents of your C:\ drive and select NEW> FOLDER. This will create a new folder. Now call this folder GIS_FOR_BIOLOGISTS by typing this into the folder name to replace what it is currently called (which will most likely be NEW FOLDER). This folder, which has the address C:\GIS_FOR_BIOLOGISTS, will be used to store all files and data for the exercises in this book.

Next, you need to download the source files for the four feature data layers from *www.gisinecology.com/GFB.htm#2* (**NOTE:** If you have already done this for exercises one or two, you do not need to download any additional data layers, as they will be in the compressed folder you downloaded at the start of those exercises). Once you have downloaded the compressed folder containing the files, make sure that you then copy all the files it contains into the folder C:\GIS_FOR_BIOLOGISTS. These feature data layers are:

1. **Mount_Mabu_Elevation.shp:** This is a line data layer which contains elevation contours for Mount Mabu and the surrounding region at 10 metre intervals. It is in the UTM Zone 37S projection/coordinate system which is based on the WGS 1984 datum.

2. **Mount_Mabu_Rainforest_Edge.shp:** This is a line data layer which represents where the outer edge of the rainforest was in 2013. It was created from a Google Earth image in the manner described in step five of exercise two (see chapter thirteen). It is in the UTM Zone 37S projection/ coordinate system which is based on the WGS 1984 datum.

3. **Mount_Mabu_Rainforest_Area.shp:** This is a polygon data layer which represents the area covered by rainforest in 2013. It was created from a Google Earth image in a manner similar to that described in step five of exercise two (see chapter thirteen). It is in the UTM Zone 37S projection/ coordinate system which is based on the WGS 1984 datum.

4. **Mount_Mabu_Summit.shp:** This is a point data layer which represents the summit of Mount Mabu. It was created from a Google Earth image in a manner similar to that described in step five of exercise two (see chapter thirteen). It is in the UTM Zone 37S projection/coordinate system which is based on the WGS 1984 datum.

Instructions For ArcGIS 10.3 Users:

Once you have all the required files downloaded into the correct folder on your computer, and you understand what is contained within each file, you can move on to creating your map. The starting point for this is a blank GIS project. To create a blank GIS project, first, start the ArcGIS software by opening the ArcMap module. When it opens, you will be presented with a window which has the heading ARCMAP – GETTING STARTED. In this window you can either select an existing GIS project to work on, or create a new one. To create a new blank GIS project, click on NEW MAPS in the directory tree on the left hand side and then select BLANK MAP in the right hand section of the window. Now, click OK at the bottom of this window. This will open a new blank GIS project. (**NOTE:** If this window does not appear, you can start a new project by clicking on FILE from the main menu bar area, and selecting NEW. When the NEW DOCUMENT window opens, select NEW MAPS and then BLANK MAP in the order outlined above.) Once you have opened your new GIS project, the first thing you need to do is save it under a new, and meaningful, name. To do this, click on FILE on the main menu bar area, and select SAVE AS. Save it as EXERCISE_THREE in the folder C:\GIS_FOR_BIOLOGISTS. Remember to save your project again at the end of each and every step (unlike exercises one and two, you will not be reminded to do this as you progress through this exercise).

STEP 1: SET THE PROJECTION AND COORDINATE SYSTEM OF YOUR DATA FRAME:

Whenever you start a new GIS project, the first thing you should do, even before you add any data layers, is select an appropriate projection/coordinate system and then set the data frame (i.e. the map you are going to add your data layers to) to use this projection/coordinate system. This ensures that you do not accidentally end up using one which is inappropriate. In addition, most problems that beginners have with GIS are caused by a failure to use an appropriate projection/coordinate system or a conflict between the projection/coordinate system used for data layers and the one being used for the data frame itself. For this exercise, you will use a pre-existing projection/coordinate system called UTM Zone 37S. This is a transverse mercator projection which is specifically designed for the area of northern Mozambique where Mount Mabu is located and it uses the WGS 1984 datum.

To set your data frame to use the UTM Zone 37S projection/coordinate system, work through the flow diagram which starts on the next page.

159

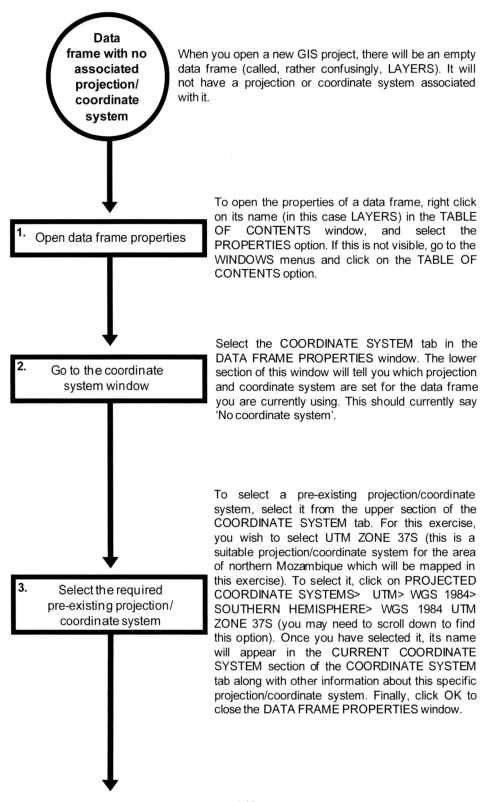

Data frame with no associated projection/ coordinate system

When you open a new GIS project, there will be an empty data frame (called, rather confusingly, LAYERS). It will not have a projection or coordinate system associated with it.

1. Open data frame properties

To open the properties of a data frame, right click on its name (in this case LAYERS) in the TABLE OF CONTENTS window, and select the PROPERTIES option. If this is not visible, go to the WINDOWS menus and click on the TABLE OF CONTENTS option.

2. Go to the coordinate system window

Select the COORDINATE SYSTEM tab in the DATA FRAME PROPERTIES window. The lower section of this window will tell you which projection and coordinate system are set for the data frame you are currently using. This should currently say 'No coordinate system'.

3. Select the required pre-existing projection/ coordinate system

To select a pre-existing projection/coordinate system, select it from the upper section of the COORDINATE SYSTEM tab. For this exercise, you wish to select UTM ZONE 37S (this is a suitable projection/coordinate system for the area of northern Mozambique which will be mapped in this exercise). To select it, click on PROJECTED COORDINATE SYSTEMS> UTM> WGS 1984> SOUTHERN HEMISPHERE> WGS 1984 UTM ZONE 37S (you may need to scroll down to find this option). Once you have selected it, its name will appear in the CURRENT COORDINATE SYSTEM section of the COORDINATE SYSTEM tab along with other information about this specific projection/coordinate system. Finally, click OK to close the DATA FRAME PROPERTIES window.

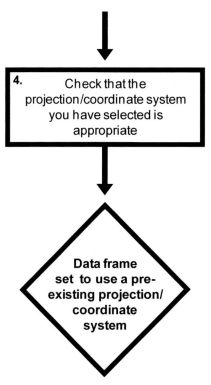

Once you have selected a projection/coordinate system, you need to check that it is appropriate. This involves examining how data layers look in it. For this exercise, this will be done in the next step by adding a series of data layers and checking that the features in them plot in the expected places and that any polygons have the expected shape.

To check that you have done this step properly, right click on the name of your data frame (LAYERS) in the TABLE OF CONTENTS window and select PROPERTIES. Click on the COORDINATE SYSTEM tab of the DATA FRAME PROPERTIES window and make sure that the contents of the CURRENT COORDINATE SYSTEM section of the window has the following text at the top of it:

WGS_1984_UTM_ZONE_37S
WKID: 32737 Authority EPSG

Projection: Transverse_Mercator
False_Easting: 500000.0
False_Northing: -10000000.0
Central_Meridian: 39.0
Scale_Factor: 0.9996
Latitude_Of_Origin: 0.0
Linear Unit: Meter (1.0)

If it does not, you will need to repeat this step until you have assigned the correct projection/coordinate system to your data frame.

STEP 2: ADD ANY EXISTING DATA LAYERS YOU NEED TO CREATE YOUR RASTER DATA LAYERS TO YOUR GIS PROJECT:

Once you have set the projection/coordinate system of your data frame, you are ready to add some existing data layers to it. For this exercise, you will add a number of existing data layers which will then be used as the basis for making the raster data layers that you wish to create. These include information about the elevation of the land in the Mount Mabu region (MOUNT_MABU_ ELEVATION), the location of the edge of the rainforest on Mount Mabu (MOUNT_MABU_RAINFOREST_EDGE) and the area of Mount Mabu covered with rainforest (MOUNT_MABU_RAINFOREST_AREA). In addition, you will add a point data layer which will mark the summit of Mount Mabu (MOUNT_MABU_SUMMIT) which will help you locate Mount Mabu in your GIS project. These existing data layers are all in the UTM Zone 37S projection which will be used for this exercise.

To add these data layers to your GIS project, work through the following flow diagram:

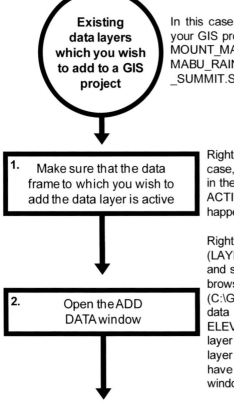

Existing data layers which you wish to add to a GIS project

In this case, the data layers which you wish to add to your GIS project are MOUNT_MABU_ELEVATION.SHP, MOUNT_MABU_RAINFOREST_AREA.SHP, MOUNT_MABU_RAINFOREST_EDGE.SHP and MOUNT_MABU _SUMMIT.SHP .

1. Make sure that the data frame to which you wish to add the data layer is active

Right click on the name of the data frame (in this case, LAYERS) which you wish to add the data to in the TABLE OF CONTENTS window, and select ACTIVATE (it will appear as if nothing has happened, but the data frame will now be active).

2. Open the ADD DATA window

Right click on the name of the data frame (LAYERS) in the TABLE OF CONTENTS window, and select ADD DATA. In the ADD DATA window, browse to the location of your data layer (C:\GIS_FOR_ BIOLOGISTS) and then select the data layer called MOUNT_MABU_ ELEVATION.SHP. Now click ADD. A line data layer will appear in your MAP window and the data layer named MOUNT_MABU_ELEVATION will have been added to the TABLE OF CONTENTS window.

3. Check the projection/ coordinate system for the newly added data layer

Whenever you add a data layer to a GIS project, you should always check that is has a projection/coordinate system assigned to it, and look at what this projection/coordinate system is. This is so that you know whether you will need to assign a projection/coordinate system to it, or transform it into a different projection/coordinate system before you can use it in your GIS project. To check the projection/coordinate system of your newly added data layer, right click on its name in the TABLE OF CONTENTS window and select PROPERTIES. In the LAYER PROPERTIES window which opens, click on the SOURCE tab and check that the projection/coordinate system listed under DATA SOURCE is WGS_1984_UTM_ Zone_37S (the same as the data frame). Now, click OK to close the LAYER PROPERTIES window. If the data layer had not been in the same projection/coordinate system as the data frame, you would have to have used the PROJECT tool to transform it into the correct projection/coordinate system. You will learn how to do this in exercise five.

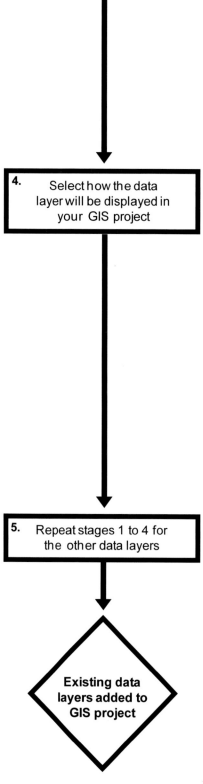

4. Select how the data layer will be displayed in your GIS project

5. Repeat stages 1 to 4 for the other data layers

Existing data layers added to GIS project

Right click on the name of the MOUNT_MABU_ ELEVATION data layer in the TABLE OF CONTENTS window, and select PROPERTIES. In the LAYER PROPERTIES window, click on the SYMBOLOGY tab. On the left hand side under show, select CATEGORIES> UNIQUE VALUES and then for VALUE FIELD select CONTOUR. This will allow you to give each different elevation contour a different colour. Next, click on ADD VALUES and then in the ADD VALUES window click on COMPLETE LIST. Once the complete list of contours has been generated, click on 500 to select it, and then scroll down until you find 1000. Now hold down the CTRL key on your keyboard and select 1000. Repeat to select 1500. Once all these contours have been selected, click OK to close the ADD VALUES window. Now, in the LAYER PROPERTIES window, untick the box next to <ALL OTHER VALUES> and then click on OK to close the LAYER PROPERTIES window. You will see that the way the MOUNT_MABU_ ELEVATION data layer is displayed in the MAP window has now changed to the new settings you have just selected. Do not worry that it looks a little strange at the moment. It will look fine once you have zoomed in on the Mount Mabu region after you have completed this step.

Repeat stages 1 to 4 for the data layers called MOUNT_MABU_RAINFOREST_AREA, MOUNT_ MABU_RAINFOREST_EDGE and MOUNT_ MABU_SUMMIT (in this order) to add them to your GIS project. For the MOUNT_MABU_ RAINFOREST_AREA, select FEATURES> SINGLE SYMBOL, and then click on the box under SYMBOL and select MACAW GREEN (the fourth colour down in the left hand column of green shades) for the FILL COLOR. For MOUNT_ MABU_RAINFOREST_EDGE select a WIDTH of 2.0, and BLACK for COLOR. For MOUNT_MABU_SUMMIT select CIRCLE 2 for the symbol, 12 for SIZE and POINSETTIA RED for COLOR.

At the end of this step, re-arrange the order of the layers in the TABLE OF CONTENTS window by clicking on the name of a data layer and dragging it to a new position until it looks like this:

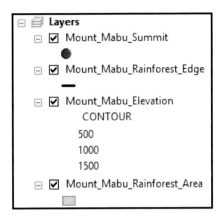

Now right click on the name of the data layer MOUNT_MABU_ RAINFOREST_AREA in the TABLE OF CONTENTS window and select ZOOM TO LAYER. The contents of your MAP window should now look like this:

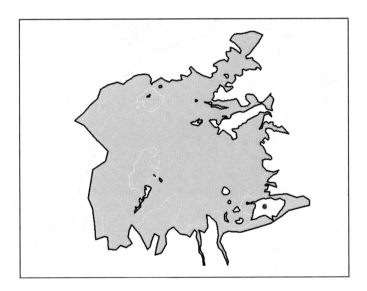

If it does not look like this, go back to step one and ensure that you have set the projection/coordinate system of your data frame correctly, and then repeat step two.

STEP 3: CREATE A RASTER DATA LAYER OF LAND ELEVATION:

Now that you have added the required existing data layers to your GIS project, you can start creating your raster data layers. The first raster data layer that you will create will be one which provides a continuous surface of land elevation information. This will essentially fill in all the gaps between the contour lines with elevation data which the GIS software will estimate from the values of the nearest contour lines and the distances between them. In the ArcGIS software package, there is a specific tool, called TOPO TO RASTER, which allows you to do this quickly and accurately. To create your raster data layer of land elevation, work through the following flow diagram (**NOTE:** If you get an error message stating that you do not have an active Spatial Analyst licence, click on CUSTOMIZE on the main menu bar, select EXTENSIONS, and then make sure that the SPATIAL ANALYST extension has been activated):

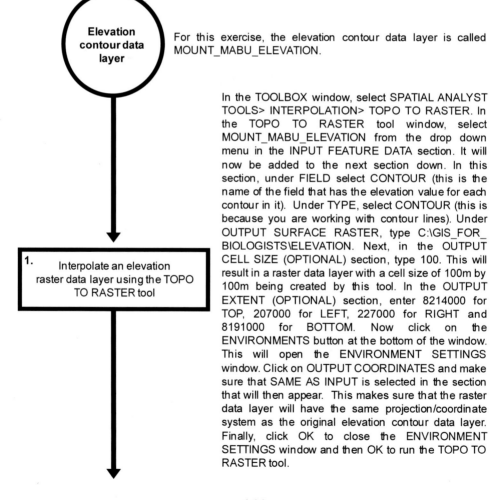

Elevation contour data layer

For this exercise, the elevation contour data layer is called MOUNT_MABU_ELEVATION.

1. Interpolate an elevation raster data layer using the TOPO TO RASTER tool

In the TOOLBOX window, select SPATIAL ANALYST TOOLS> INTERPOLATION> TOPO TO RASTER. In the TOPO TO RASTER tool window, select MOUNT_MABU_ELEVATION from the drop down menu in the INPUT FEATURE DATA section. It will now be added to the next section down. In this section, under FIELD select CONTOUR (this is the name of the field that has the elevation value for each contour in it). Under TYPE, select CONTOUR (this is because you are working with contour lines). Under OUTPUT SURFACE RASTER, type C:\GIS_FOR_ BIOLOGISTS\ELEVATION. Next, in the OUTPUT CELL SIZE (OPTIONAL) section, type 100. This will result in a raster data layer with a cell size of 100m by 100m being created by this tool. In the OUTPUT EXTENT (OPTIONAL) section, enter 8214000 for TOP, 207000 for LEFT, 227000 for RIGHT and 8191000 for BOTTOM. Now click on the ENVIRONMENTS button at the bottom of the window. This will open the ENVIRONMENT SETTINGS window. Click on OUTPUT COORDINATES and make sure that SAME AS INPUT is selected in the section that will then appear. This makes sure that the raster data layer will have the same projection/coordinate system as the original elevation contour data layer. Finally, click OK to close the ENVIRONMENT SETTINGS window and then OK to run the TOPO TO RASTER tool.

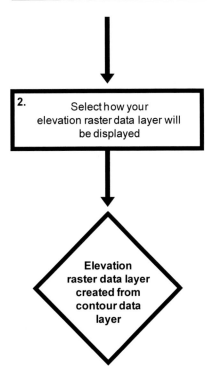

Right click on the name of the ELEVATION data layer in the TABLE OF CONTENTS window, and select PROPERTIES. In the LAYER PROPERTIES window, click on the SYMBOLOGY tab. On the left hand side, under SHOW, select STRETCHED. Next, beside COLOR RAMP, select a scale that grades from black on the left to white on the right. Finally, click OK to close the LAYER PROPERTIES window.

At the end of this step, your TABLE OF CONTENTS window should look like this (**NOTE:** The exact range of values you get for your ELEVATION raster data layer may vary slightly depending on the exact version of the ArcGIS software you are using):

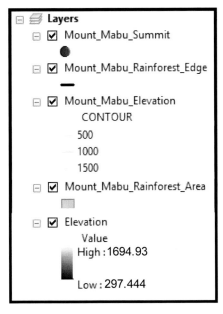

Now, click on the box next to MOUNT_MABU_RAINFOREST_AREA in the TABLE OF CONTENTS window so that it is no longer displayed in the MAP window. The contents of your MAP window should now look like this:

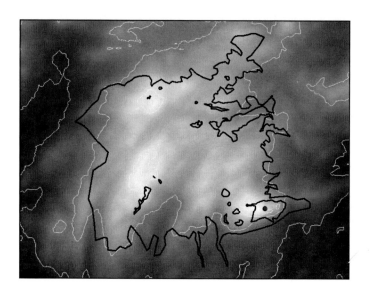

STEP 4: CREATE A RASTER DATA LAYER OF SLOPE:

Once you have created your raster data layer of elevation from your elevation contour data layer, you have effectively created what is known as a digital elevation model (or DEM for short). DEMs are very useful, not only because they provide you with information on the elevation of different locations in the area covered by your GIS project, but also because they can be used to generate additional raster data layers which provide information on other environmental variables which might influence the distribution of plants and animals, such as how steeply the land slopes, the amount of direct sunlight that each location receives (something known as hillshade), how water will flow across the local terrain (known as flow), the direction that the slope faces (known as aspect), how the terrain varies between neighbouring cells in the DEM (known as ruggedness) and whether specific objects, such as wind turbines, can be seen from different locations within a particular area (known as a viewshed or zone of theoretical visibility). All of these can be derived from a DEM using tools found in most GIS software packages. For this exercise, you will derive two such data layers: slope (in this step) and hillshade (in step 5). The raster data layer of slope can be derived from the DEM by working through the flow diagram which starts on the next page.

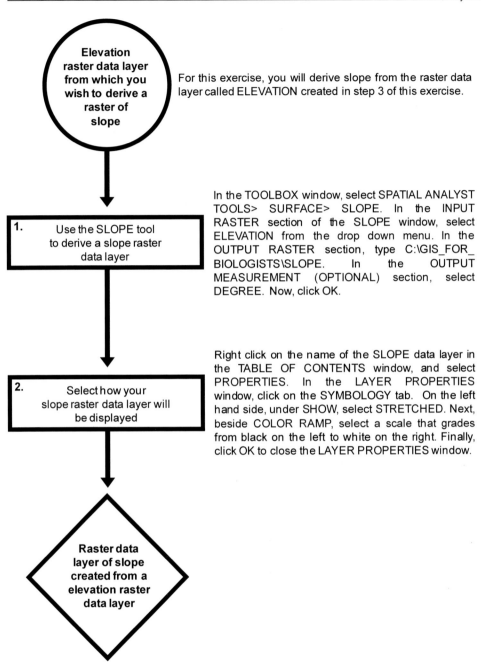

For this exercise, you will derive slope from the raster data layer called ELEVATION created in step 3 of this exercise.

In the TOOLBOX window, select SPATIAL ANALYST TOOLS> SURFACE> SLOPE. In the INPUT RASTER section of the SLOPE window, select ELEVATION from the drop down menu. In the OUTPUT RASTER section, type C:\GIS_FOR_ BIOLOGISTS\SLOPE. In the OUTPUT MEASUREMENT (OPTIONAL) section, select DEGREE. Now, click OK.

Right click on the name of the SLOPE data layer in the TABLE OF CONTENTS window, and select PROPERTIES. In the LAYER PROPERTIES window, click on the SYMBOLOGY tab. On the left hand side, under SHOW, select STRETCHED. Next, beside COLOR RAMP, select a scale that grades from black on the left to white on the right. Finally, click OK to close the LAYER PROPERTIES window.

At the end of this step, your TABLE OF CONTENTS window should look like the image at the top of the next page (**NOTE:** The range of values you get for your SLOPE raster data layer may vary slightly depending on the exact version of the ArcGIS software you are using).

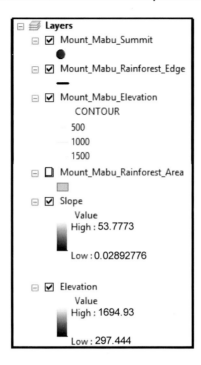

While the contents of your MAP window should look like this:

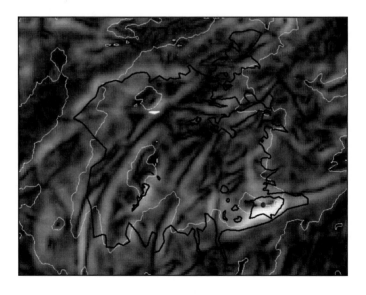

STEP 5: CREATE A RASTER DATA LAYER OF HILLSHADE:

Hillshade uses information about the surrounding land elevation to work out how much direct sunlight each grid cell within a raster data layer will receive when the sun is at a specific position in the sky. The position of the sun will be determined by two components: Time of day, which provides a measure of how high the sun is above the horizon, and time of year, which provides information about the direction the sun will be shining from at that time of day. Together, these give the altitude and azimuth of the sun which will be used to derive the hillshade raster data layer from the DEM. For this exercise, you will use an altitude of 30 degrees and an azimuth of 135 degrees. This will provide a measure of now much sunlight each grid cell gets in the early morning around the southern summer solstice (see note on page 191 for information on how to obtain values for other locations, dates and times of day). This can be done by working through the following flow diagram:

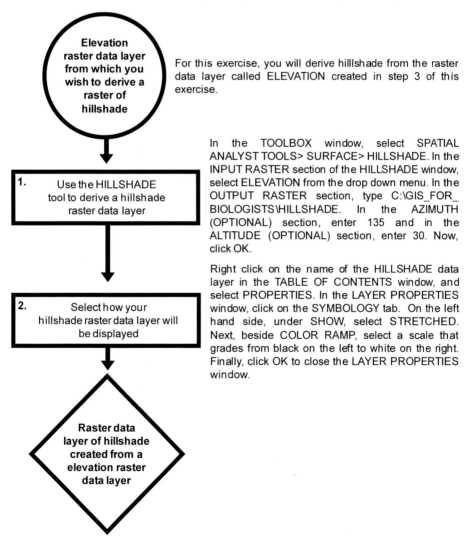

Elevation raster data layer from which you wish to derive a raster of hillshade

For this exercise, you will derive hilllshade from the raster data layer called ELEVATION created in step 3 of this exercise.

1. Use the HILLSHADE tool to derive a hillshade raster data layer

In the TOOLBOX window, select SPATIAL ANALYST TOOLS> SURFACE> HILLSHADE. In the INPUT RASTER section of the HILLSHADE window, select ELEVATION from the drop down menu. In the OUTPUT RASTER section, type C:\GIS_FOR_ BIOLOGISTS\HILLSHADE. In the AZIMUTH (OPTIONAL) section, enter 135 and in the ALTITUDE (OPTIONAL) section, enter 30. Now, click OK.

2. Select how your hillshade raster data layer will be displayed

Right click on the name of the HILLSHADE data layer in the TABLE OF CONTENTS window, and select PROPERTIES. In the LAYER PROPERTIES window, click on the SYMBOLOGY tab. On the left hand side, under SHOW, select STRETCHED. Next, beside COLOR RAMP, select a scale that grades from black on the left to white on the right. Finally, click OK to close the LAYER PROPERTIES window.

Raster data layer of hillshade created from a elevation raster data layer

At the end of this step, your TABLE OF CONTENTS window should look like the image at the top of the next page.

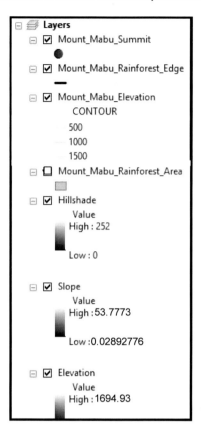

While the contents of your MAP window should look like this:

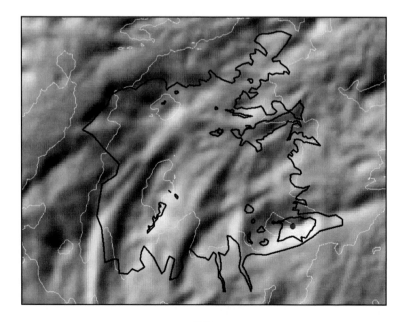

STEP 6: CREATE A RASTER DATA LAYER OF DISTANCE TO THE EDGE OF THE RAINFOREST:

So far, the raster data layers of environmental variables have all been derived from information on the elevation of the land around Mount Mabu. However, raster data layers can also be derived from many other types of environmental information. As a demonstration of this, in this step, a raster data layer will be created which measures the distance of each grid cell within the rainforest from the forest edge. This information can be used to identify areas where you are likely to find species which prefer to occur close to the edges of the forest, or near breaks in the forest canopy, and areas where you are likely to find species that only occur away from the forest edge. This raster data layer will be created in two stages. Firstly, a raster data layer will be created with a tool called EUCLIDEAN DISTANCE which will calculate the distance that each grid cell is from the line marking the edge of the forest (MOUNT_MABU_RAINFOREST_EDGE). When this data layer is made, you will see that it calculates distance values for all cells in the raster data layer and not just the ones which are inside of the area of forest. To remove the non-forest cells, in the second stage, the polygon representing the area of forest in the data layer MOUNT_MABU_RAINFOREST_AREA is used to mask the distance raster data layer so that only the distance values inside the forest are retained. To create this raster data layer, work through the following flow diagram:

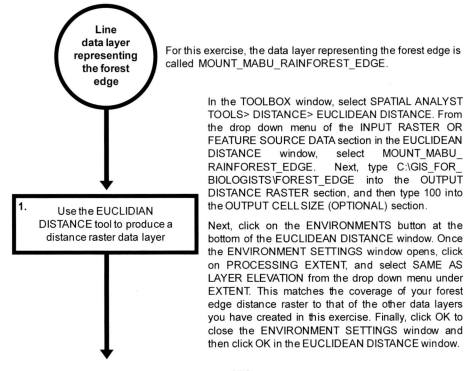

Line data layer representing the forest edge

For this exercise, the data layer representing the forest edge is called MOUNT_MABU_RAINFOREST_EDGE.

1. Use the EUCLIDIAN DISTANCE tool to produce a distance raster data layer

In the TOOLBOX window, select SPATIAL ANALYST TOOLS> DISTANCE> EUCLIDEAN DISTANCE. From the drop down menu of the INPUT RASTER OR FEATURE SOURCE DATA section in the EUCLIDEAN DISTANCE window, select MOUNT_MABU_RAINFOREST_EDGE. Next, type C:\GIS_FOR_BIOLOGISTS\FOREST_EDGE into the OUTPUT DISTANCE RASTER section, and then type 100 into the OUTPUT CELL SIZE (OPTIONAL) section.

Next, click on the ENVIRONMENTS button at the bottom of the EUCLIDEAN DISTANCE window. Once the ENVIRONMENT SETTINGS window opens, click on PROCESSING EXTENT, and select SAME AS LAYER ELEVATION from the drop down menu under EXTENT. This matches the coverage of your forest edge distance raster to that of the other data layers you have created in this exercise. Finally, click OK to close the ENVIRONMENT SETTINGS window and then click OK in the EUCLIDEAN DISTANCE window.

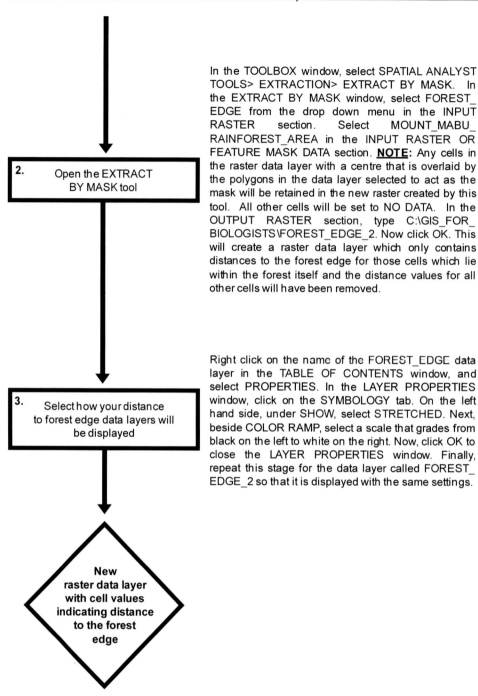

2. Open the EXTRACT BY MASK tool

In the TOOLBOX window, select SPATIAL ANALYST TOOLS> EXTRACTION> EXTRACT BY MASK. In the EXTRACT BY MASK window, select FOREST_ EDGE from the drop down menu in the INPUT RASTER section. Select MOUNT_MABU_ RAINFOREST_AREA in the INPUT RASTER OR FEATURE MASK DATA section. **NOTE:** Any cells in the raster data layer with a centre that is overlaid by the polygons in the data layer selected to act as the mask will be retained in the new raster created by this tool. All other cells will be set to NO DATA. In the OUTPUT RASTER section, type C:\GIS_FOR_ BIOLOGISTS\FOREST_EDGE_2. Now click OK. This will create a raster data layer which only contains distances to the forest edge for those cells which lie within the forest itself and the distance values for all other cells will have been removed.

3. Select how your distance to forest edge data layers will be displayed

Right click on the name of the FOREST_EDGE data layer in the TABLE OF CONTENTS window, and select PROPERTIES. In the LAYER PROPERTIES window, click on the SYMBOLOGY tab. On the left hand side, under SHOW, select STRETCHED. Next, beside COLOR RAMP, select a scale that grades from black on the left to white on the right. Now, click OK to close the LAYER PROPERTIES window. Finally, repeat this stage for the data layer called FOREST_ EDGE_2 so that it is displayed with the same settings.

New raster data layer with cell values indicating distance to the forest edge

At the end of this step, your TABLE OF CONTENTS window should look like the image at the top of the next page.

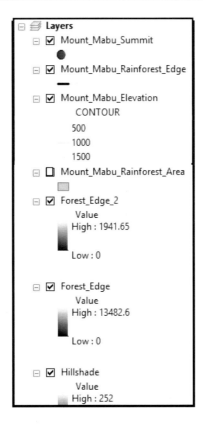

In the TABLE OF CONTENTS window, untick the box next to FOREST_EDGE_2 so that it is no longer displayed in the MAP window. At this point, the contents of your MAP window should look like this:

Now, in the TABLE OF CONTENTS window, click on the box next to FOREST_EDGE_2 so that it is displayed again. Then, untick the boxes next to all the other raster data layers so that they are no longer displayed in the MAP window. This will allow you to see the effect of masking the FOREST_EDGE raster data layer to remove the cells which are outside of the rainforest area. The contents of your MAP window should look like this:

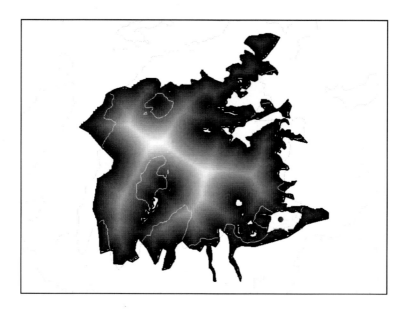

You can now remove the data layer called FOREST_EDGE from your GIS project as it is no longer needed. This is done by right-clicking on its name in the TABLE OF CONTENTS window and selecting REMOVE. The final FOREST_EDGE_2 raster data layer can be used to identify where species that prefer to occur well away from the edge of the forest are likely to occur, as well as where species which inhabit the forest edge are likely to be found.

STEP 7: CREATE A RASTER DATA LAYER REPRESENTING THE EXTENT OF THE RAINFOREST AREA ON MOUNT MABU IN 2013:

Up to this point in this exercise, you have been working with environmental data layers and using interpolation tools, such as TOPO TO RASTER, to produce continuous surfaces when converting feature data layers to raster data layers. However, there is another way to create raster data layers from features, and this is to use what is known as a 'convert to grid' approach. In the 'interpolate grid' approach used for the environmental raster data layers, all the cells within the new raster data layer have values in them, with the tool interpolating values for any grid cells that do not contain a feature from the original feature data layer (this was the reason that the first raster data layer of the distance to the forest edge had to be masked to remove the distance values from the cells which fell outside the area of rainforest). In contrast, in the 'convert to grid' approach, a grid cell only receives a value if it contains a feature from the feature data layer. All those cells that do no contain a feature are given a 'no data' value and this means that they are treated as empty cells in the final raster data layer.

To demonstrate the 'convert to grid' approach, you will create a raster data layer where cells which fall within the area covered by rainforest in 2013 have a value, and all other cells will be classified as 'no data'. This raster data layer will be made from the MOUNT_MABU_RAINFOREST_AREA polygon data layer which was originally created using the Google Earth user interface from an image taken in 2013, and which has been provided for this exercise. This means that this raster data layer will represent a baseline of forest coverage to which similar raster data layers from future images can be compared. Grid cells which have changed from being classified as containing rainforest to not containing rainforest between time periods can be used to identify not only whether deforestation has taken place, but in which grid cells it has occurred. Thus, this step creates a raster data layer which can be used to monitor the Mount Mabu rainforest area for potential changes which could indicate detrimental changes to this habitat, and so the organisms which occur there. In addition, by comparing these cells to the raster data layers of environmental variables, you would be able to tell whether changes to the rainforest coverage are concentrated in specific types of habitats, such as those close to the edge, those at lower altitudes or those in areas with gentle slopes which are easier to access and log.

Thus, the raster data layer of forest coverage which will be generated in this step could be used as part of a monitoring programme, with new raster data layers being created using the same methods whenever new images from this part of the world are available through the Google Earth user interface, or from any other suitable sources. This means that it is possible to monitor impacts of deforestation on Mount Mabu without having to necessarily visit the local area, allowing it to be conducted at relatively low cost.

To create this raster data layer of rainforest cover, work through the flow diagram which starts on the next page.

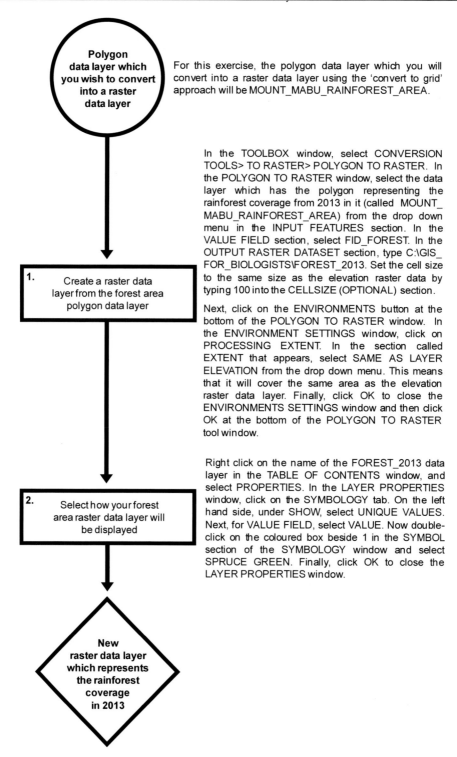

Polygon data layer which you wish to convert into a raster data layer

For this exercise, the polygon data layer which you will convert into a raster data layer using the 'convert to grid' approach will be MOUNT_MABU_RAINFOREST_AREA.

1. Create a raster data layer from the forest area polygon data layer

In the TOOLBOX window, select CONVERSION TOOLS> TO RASTER> POLYGON TO RASTER. In the POLYGON TO RASTER window, select the data layer which has the polygon representing the rainforest coverage from 2013 in it (called MOUNT_ MABU_RAINFOREST_AREA) from the drop down menu in the INPUT FEATURES section. In the VALUE FIELD section, select FID_FOREST. In the OUTPUT RASTER DATASET section, type C:\GIS_ FOR_BIOLOGISTS\FOREST_2013. Set the cell size to the same size as the elevation raster data by typing 100 into the CELLSIZE (OPTIONAL) section.

Next, click on the ENVIRONMENTS button at the bottom of the POLYGON TO RASTER window. In the ENVIRONMENT SETTINGS window, click on PROCESSING EXTENT. In the section called EXTENT that appears, select SAME AS LAYER ELEVATION from the drop down menu. This means that it will cover the same area as the elevation raster data layer. Finally, click OK to close the ENVIRONMENTS SETTINGS window and then click OK at the bottom of the POLYGON TO RASTER tool window.

2. Select how your forest area raster data layer will be displayed

Right click on the name of the FOREST_2013 data layer in the TABLE OF CONTENTS window, and select PROPERTIES. In the LAYER PROPERTIES window, click on the SYMBOLOGY tab. On the left hand side, under SHOW, select UNIQUE VALUES. Next, for VALUE FIELD, select VALUE. Now double-click on the coloured box beside 1 in the SYMBOL section of the SYMBOLOGY window and select SPRUCE GREEN. Finally, click OK to close the LAYER PROPERTIES window.

New raster data layer which represents the rainforest coverage in 2013

At the end of this step, your TABLE OF CONTENTS window should look like the image at the top of the next page.

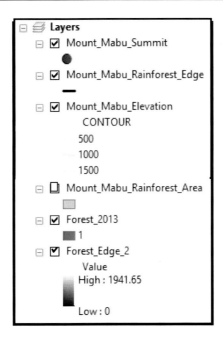

Now click on the boxes next to the names of all the raster data layers in the TABLE OF CONTENTS window so that the only raster data layer which is displayed is FOREST_2013. The contents of your MAP window should now look like this:

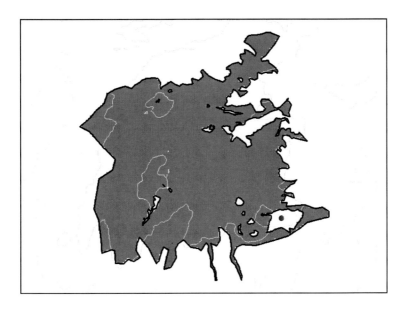

This is the end of exercise three. Now that you have completed it, you know how to create new raster data layers using two different approaches (the 'convert to grid' approach used to create the raster data layer of the forest coverage, and the 'interpolate grid' approach used to create the elevation and distance to forest edge raster data layers). You also now know how to derive new data layers from existing raster data layers (such as the slope and the hillshade raster data layers, which were derived from the elevation raster data layer) and how to mask a raster data layer to remove unwanted values (as was done with the forest edge raster data layer). Together, these represent the main methods that biologists need to know to create new raster data layers in a GIS project. However, one further method (using a raster calculator tool) will be introduced in exercise six in chapter seventeen.

Instructions For QGIS 2.8.3 Users:

Once you have all the required files downloaded into the correct folder on your computer, and you understand what is contained within each file, you can move on to creating your GIS project for this exercise. The starting point for this is a blank GIS project. To create a blank GIS project, first open QGIS. Once it is open, click on the PROJECT menu and select SAVE AS. In the window which opens, save your GIS project as EXERCISE_THREE in the folder C:\GIS_FOR_ BIOLOGISTS. Remember to save your project again at the end of each and every step (unlike exercises one and two, you will not be reminded to do this as you progress through this exercise).

STEP 1: SET THE PROJECTION AND COORDINATE SYSTEM OF YOUR DATA FRAME:

Whenever you start a new GIS project, the first thing you should do, even before you add any data layers, is select an appropriate projection/coordinate system and then set the data frame (i.e. the map you are going to add your data layers to) to use this projection/coordinate system. This ensures that you do not accidently end up using one which is inappropriate. In addition, most problems that beginners have with GIS are caused by a failure to use an appropriate projection/coordinate system or a conflict between the projection/coordinate system used for data layers and the one being used for the data frame itself. For this exercise, you will use a pre-existing projection/ coordinate system called Universal Transverse Mercator (UTM) Zone 37S. This is a transverse mercator projection which is specifically designed for the area of northern Mozambique where Mount Mabu is located and it uses the WGS 1984 datum. To set your data frame to use the UTM Zone 37S projection/coordinate system, work through the flow diagram which starts on the next page.

Data frame with no associated projection and coordinate system

When you open a new GIS project, there will be an empty data frame. It will not have a projection or coordinate system associated with it.

In the main QGIS window, click on the CRS STATUS button in the bottom right hand corner (it is square with a dark circular design on it). This will open the PROJECT PROPERTIES CRS window. Click on the box next to ENABLE 'ON THE FLY' CRS TRANSFORMATION so that a cross appears in it. Next, type the name of the projection/ coordinate system you want to use into the FILTER section. In this case, it will be WGS 84 / UTM ZONE 37S – **NOTE:** You will have to enter this name exactly as written, including the spaces, in order to be able to find it. In the COORDINATE REFERENCE SYSTEMS OF THE WORLD section, click on WGS 84 / UTM ZONE 37S under PROJECTED COORDINATE SYSTEMS> UNIVERSAL TRANSVERSE MERCATOR (UTM). This will add this to the SELECTED CRS section further down the window. Now click OK to close the PROJECT PROPERTIES CRS window.

1.   Set your data frame to use a pre-existing projection/ coordinate system

Once you have selected a projection/ coordinate system, you need to check that it is appropriate. This involves examining how data layers look in it. For this exercise, this will be done in the next step by adding a series of data layers and checking that the features in them plot in the expected places and that any polygons have the expected shape.

2.   Check that the projection/coordinate system you have selected is appropriate

Data frame set to use a pre-existing projection/ coordinate system

To check that you have done this step properly, click on PROJECT on the main menu bar and select PROJECT PROPERTIES. In the PROJECT PROPERTIES window which will open, click on the CRS tab on the left hand side and make sure that the contents of the SELECTED CRS section has the following text in it:

WGS 84 / UTM ZONE 37S

Underneath this it should say:

   +proj=utm +zone=37 +south +datum=WGS84 +units=m +no_defs

This is known as the Proj.4 string, which is used to define the characteristics of the projection/coordinate system.

If it does, click OK to close the PROJECT PROPERTIES window. If it does not, you will need to repeat this step until you have assigned the correct projection/coordinate system to your data frame.

STEP 2: ADD ANY EXISTING DATA LAYERS YOU NEED TO CREATE YOUR RASTER DATA LAYERS TO YOUR GIS PROJECT:

Once you have set the projection/coordinate system of your data frame, you are ready to add some existing data layers to it. For this exercise, you will add a number of existing data layers which will be used as the basis for making the raster data layers which you wish to create. This includes information about the elevation of the land in the Mount Mabu region (MOUNT_MABU_ ELEVATION), the location where the edge of the rainforest is located on Mount Mabu (MOUNT_MABU_RAINFOREST_EDGE) and the area of Mount Mabu covered with rainforest (MOUNT_MABU_RAINFOREST_ AREA). In addition, you will add a point data layer which will mark the summit of Mount Mabu (MOUNT_MABU_SUMMIT) which will help you locate Mount Mabu in your GIS project. These existing data layers are all in the UTM Zone 37S projection/coordinate system which will be used for this exercise.

To add these data layers to your GIS project, work through the flow diagram which starts on the next page.

Existing
data layers
which you wish
to add to a GIS
project

In this case, the data layers which you wish to add to your GIS project are MOUNT_MABU_ELEVATION.SHP, MOUNT_MABU_RAINFOREST_AREA.SHP, MOUNT_ MABU_RAINFOREST_EDGE.SHP and MOUNT_MABU _SUMMIT.SHP.

1. Open the ADD
VECTOR LAYER window

On the main menu bar, click on LAYER and select ADD LAYER> ADD VECTOR LAYER. In the ADD VECTOR LAYER window, browse to the location of your data layer (C:\GIS_FOR_BIOLOGISTS) and click on the section in the bottom right hand corner of the OPEN window (where it currently says ALL FILES (*).(*)) and select ESRI SHAPEFILES (*.shp, .SHP). Now, select the data layer called MOUNT_MABU_ELEVATION.SHP. Next, click OPEN in the browse window and then OPEN in the ADD VECTOR LAYER window.

2. Check the projection/
coordinate system for the newly
added data layer

Whenever you add a data layer to a GIS project, you should always check that it has a projection/coordinate system assigned to it, and look at what this projection/coordinate system is. This is so that you know whether you will need to assign a projection/coordinate system to it, or transform it into a different projection/coordinate system before you can use it in your GIS project. To check the projection/coordinate system of your newly added data layer, right click on its name in the TABLE OF CONTENTS window and select PROPERTIES. In the LAYER PROPERTIES window which opens, click on the GENERAL tab on the left hand side and check that there is a projection/coordinate system listed in the COORDINATE REFERENCE SYSTEM window. For the MOUNT_MABU_ELEVATION data layer this should say :

EPSG:32737 – WGS 84 / UTM ZONE 37S

This tells you that this data layer is in the UTM Zone 37S projection based on the WGS 1984 datum. Now click OK to close the LAYER PROPERTIES window.

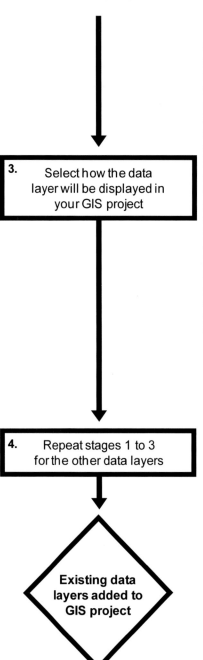

3. Select how the data layer will be displayed in your GIS project

Right click on the name of the MOUNT_MABU_ ELEVATION data layer in the TABLE OF CONTENTS window, and select PROPERTIES. In the LAYER PROPERTIES window, click on the STYLE tab on the left hand side. In the top left hand corner of the STYLE tab, select CATEGORIZED from the drop down menu in the section which currently says SINGLE SYMBOL. For COLUMN, select CONTOUR from the drop down menu. Now click on ADD. This will add a line symbol in the section directly above this. In this section, double click beside this line under VALUE and enter the number 0. Double click under LEGEND and enter 0 here too. This tells QGIS to display the 0 elevation contour using this symbol. Next, click ADD again to add another line. Next to this, enter 500 for VALUE and 500 for LEGEND. Repeat this for 1000, 1500, 2000 and 2500. This will mean that only contours with these elevations will be displayed in the MAP window. Now click OK to close the LAYER PROPERTIES window.

4. Repeat stages 1 to 3 for the other data layers

Existing data layers added to GIS project

Now right click on the name of the data layer MOUNT_MABU_ RAINFOREST_AREA in the TABLE OF CONTENTS window and select ZOOM TO LAYER. Then click on its name again and drag it down to the bottom of the list of data layers in the TABLE OF CONTENTS window.

At the end of this step, your TABLE OF CONTENTS window should look like this:

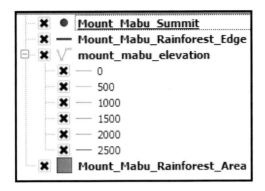

While the contents of your MAP window should look like this:

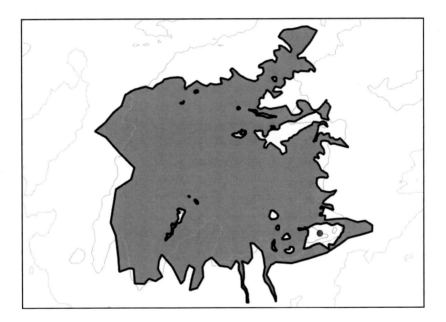

If it does not look like this, go back to step one and ensure that you have set the projection/coordinate system of your data frame correctly, and then repeat step two.

STEP 3: CREATE A RASTER DATA LAYER OF LAND ELEVATION:

Now that you have added the required existing data layers to your GIS project, you can start creating your raster data layers. The first raster data layer that you will create will be one which provides a continuous surface of land elevation information. This will essentially fill in all the gaps between the contour lines with elevation data which it will estimate from the values of the nearest contour lines and the distances between them. In the QGIS software package, there is a specific tool, called R.SURF.CONTOUR, which allows you to do this quickly and accurately. To create your raster data layer of land elevation, work through the following flow diagram (**NOTE:** To access the GRASS tools used in this step, you will need to have set your TOOLBOX window to ADVANCED INTERFACE – see page 69 for more information):

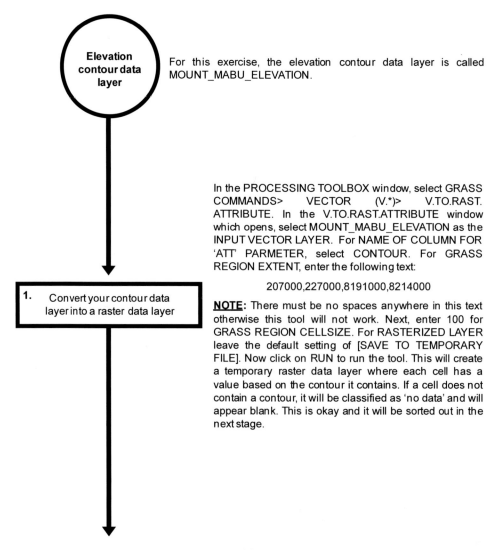

Elevation contour data layer

For this exercise, the elevation contour data layer is called MOUNT_MABU_ELEVATION.

1. Convert your contour data layer into a raster data layer

In the PROCESSING TOOLBOX window, select GRASS COMMANDS> VECTOR (V.*)> V.TO.RAST. ATTRIBUTE. In the V.TO.RAST.ATTRIBUTE window which opens, select MOUNT_MABU_ELEVATION as the INPUT VECTOR LAYER. For NAME OF COLUMN FOR 'ATT' PARMETER, select CONTOUR. For GRASS REGION EXTENT, enter the following text:

207000,227000,8191000,8214000

NOTE: There must be no spaces anywhere in this text otherwise this tool will not work. Next, enter 100 for GRASS REGION CELLSIZE. For RASTERIZED LAYER leave the default setting of [SAVE TO TEMPORARY FILE]. Now click on RUN to run the tool. This will create a temporary raster data layer where each cell has a value based on the contour it contains. If a cell does not contain a contour, it will be classified as 'no data' and will appear blank. This is okay and it will be sorted out in the next stage.

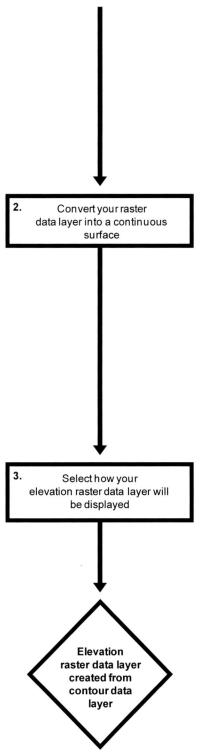

2. Convert your raster data layer into a continuous surface

3. Select how your elevation raster data layer will be displayed

Elevation raster data layer created from contour data layer

Now that you have converted your contours into a raster data layer, you are ready to convert them into a continuous surface that has a value for elevation for every grid cell in the raster data layer. In the PROCESSING TOOLBOX window, select GRASS COMMANDS> RASTER (R.*)> R.SURF.CONTOUR. In the R.SURF.CONTOUR window which opens, select the data layer you just created (RASTERIZED LAYER) as the RASTER LAYER WITH RASTERIZED CONTOURS. For GRASS REGION EXTENT, enter the following text:

207000,227000,8191000,8214000

NOTE: There must be no spaces anywhere in this text otherwise this tool will not work. Next, enter 100 for GRASS REGION CELLSIZE. For OUTPUT RASTER LAYER (this may be called DTM FROM CONTOURS in some versions of QGIS), type C:/GIS_FOR_ BIOLOGISTS/ELEVATION.TIF. Now click on RUN to run the tool. Once it has finished running, right click on the data layer which this tool creates (called either OUTPUT RASTER LAYER or DTM FROM CONTOURS) in the TABLE OF CONTENTS window and select RENAME. Once its existing name has been selected, type ELEVATION so that it replaces it and then press the ENTER key on your keyboard. Finally, right click on the temporary raster data layer named RASTERIZED LAYER in the TABLE OF CONTENTS window and select REMOVE.

Right click on the name of the ELEVATION data layer in the TABLE OF CONTENTS window, and select PROPERTIES. In the LAYER PROPERTIES window, click on the STYLE tab on the left hand side. In the top left hand corner of the STYLE tab, select SINGLEBAND GRAY from the drop down menu next to RENDER TYPE. For COLOR GRADIENT, select BLACK TO WHITE and for CONTRAST ENHANCEMENT, select STRETCH TO MINMAX. Finally, click OK to close the LAYER PROPERTIES window.

At the end of this step, change the position of the data layer called ELEVATION in your TABLE OF CONTENTS window by clicking on it and dragging it until it is between the MOUNT_MABU_ELEVATION and the MOUNT_MABU_RAINFOREST_AREA data layers (**NOTE:** You only need to do this if it is not already in that position). Your TABLE OF CONTENTS window should now look like this:

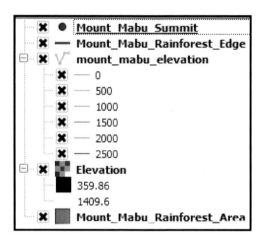

While the contents of your MAP window should look like this:

STEP 4: CREATE A RASTER DATA LAYER OF SLOPE:

Once you have created your raster data layer of elevation from your elevation contour data layer, you have effectively created what is known as a digital elevation model (or DEM for short). DEMs are very useful, not only because they provide you with information on the elevation of different locations in the area covered by your GIS project, but also because they can be used to generate additional raster data layers which provide information on other environmental variables which might influence the distribution of plants and animals, such as how steeply the land slopes, the amount of direct sunlight that each location receives (something known as hillshade), how water will flow across the local terrain (known as flow), the direction that the slope faces (known as aspect), how the terrain varies between neighbouring cells in the DEM (known as ruggedness) and whether specific objects, such as wind turbines, can be seen from different locations within a particular area (known as a viewshed or zone of theoretical visibility). All of these can be derived from a DEM using tools found in most GIS software packages. For this exercise, you will derive two such data layers: slope (in this step) and hillshade (in step 5). The raster data layer of slope can be derived from the DEM by working through the following flow diagram:

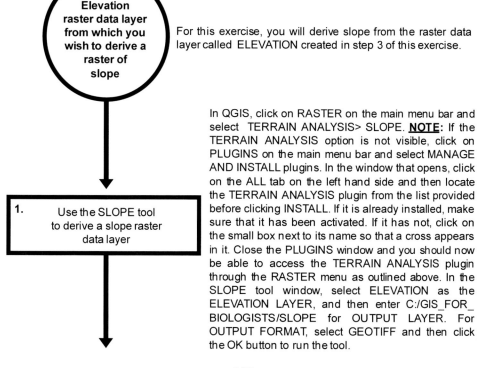

Elevation raster data layer from which you wish to derive a raster of slope

For this exercise, you will derive slope from the raster data layer called ELEVATION created in step 3 of this exercise.

1. Use the SLOPE tool to derive a slope raster data layer

In QGIS, click on RASTER on the main menu bar and select TERRAIN ANALYSIS> SLOPE. **NOTE:** If the TERRAIN ANALYSIS option is not visible, click on PLUGINS on the main menu bar and select MANAGE AND INSTALL plugins. In the window that opens, click on the ALL tab on the left hand side and then locate the TERRAIN ANALYSIS plugin from the list provided before clicking INSTALL. If it is already installed, make sure that it has been activated. If it has not, click on the small box next to its name so that a cross appears in it. Close the PLUGINS window and you should now be able to access the TERRAIN ANALYSIS plugin through the RASTER menu as outlined above. In the SLOPE tool window, select ELEVATION as the ELEVATION LAYER, and then enter C:/GIS_FOR_ BIOLOGISTS/SLOPE for OUTPUT LAYER. For OUTPUT FORMAT, select GEOTIFF and then click the OK button to run the tool.

189

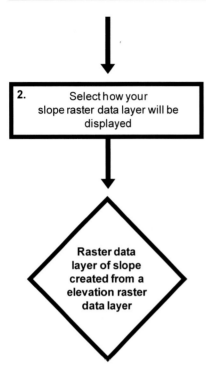

Right click on the name of the SLOPE data layer in the TABLE OF CONTENTS window, and select PROPERTIES. In the LAYER PROPERTIES window, click on the STYLE tab on the left hand side. In the top left hand corner of the STYLE tab, select SINGLEBAND GRAY from the drop down menu next to RENDER TYPE. For COLOR GRADIENT, select BLACK TO WHITE and for CONTRAST ENHANCEMENT, select STRETCH TO MINMAX. Finally, click OK to close the LAYER PROPERTIES window.

At the end of this step, change the position of the data layer called SLOPE in your TABLE OF CONTENTS window by clicking on it and dragging it until it is between the MOUNT_MABU_ELEVATION and the ELEVATION data layers. Your TABLE OF CONTENTS window should now look like this:

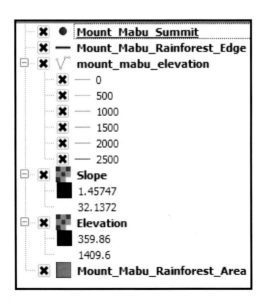

While the contents of your MAP window should look like this:

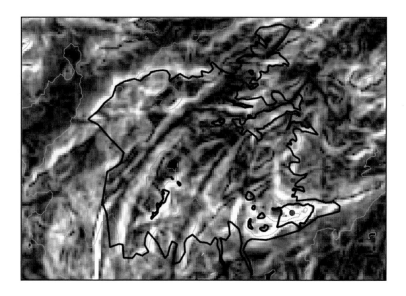

STEP 5: CREATE A RASTER DATA LAYER OF HILLSHADE:

Hillshade uses information about the surrounding land elevation to work out how much direct sunlight each grid cell within a raster data layer will receive when the sun is at a specific position in the sky. The position of the sun will be determined by two components: Time of day, which provides a measure of how high the sun is above the horizon, and time of year, which provides information about the direction the sun will be shining from at that time of day. Together, these give the altitude and azimuth of the sun which will be used to derive the hillshade raster data layer from the DEM. For this exercise, you will use an altitude of 30 degrees and an azimuth of 135 degrees. This will provide a measure of now much sunlight each grid cell gets in the early morning around the time of the southern summer solstice. This can be done by working through the flow diagram on the next page.

NOTE: If you wish to work out what altitude and azimuth to use to calculate hillshade for any specific location, date and time of day, you can use the information on this website to help you: *www.pveducation.org/pvcdrom/properties-of-sunlight/sun-position-calculator.*

191

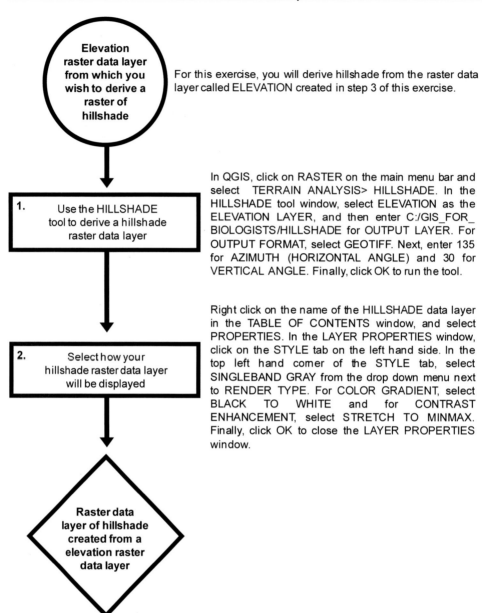

Elevation raster data layer from which you wish to derive a raster of hillshade

For this exercise, you will derive hillshade from the raster data layer called ELEVATION created in step 3 of this exercise.

1. Use the HILLSHADE tool to derive a hillshade raster data layer

In QGIS, click on RASTER on the main menu bar and select TERRAIN ANALYSIS> HILLSHADE. In the HILLSHADE tool window, select ELEVATION as the ELEVATION LAYER, and then enter C:/GIS_FOR_ BIOLOGISTS/HILLSHADE for OUTPUT LAYER. For OUTPUT FORMAT, select GEOTIFF. Next, enter 135 for AZIMUTH (HORIZONTAL ANGLE) and 30 for VERTICAL ANGLE. Finally, click OK to run the tool.

2. Select how your hillshade raster data layer will be displayed

Right click on the name of the HILLSHADE data layer in the TABLE OF CONTENTS window, and select PROPERTIES. In the LAYER PROPERTIES window, click on the STYLE tab on the left hand side. In the top left hand corner of the STYLE tab, select SINGLEBAND GRAY from the drop down menu next to RENDER TYPE. For COLOR GRADIENT, select BLACK TO WHITE and for CONTRAST ENHANCEMENT, select STRETCH TO MINMAX. Finally, click OK to close the LAYER PROPERTIES window.

Raster data layer of hillshade created from a elevation raster data layer

At the end of this step, change the position of the data layer called HILLSHADE in your TABLE OF CONTENTS window by clicking on it and dragging it until it is between the MOUNT_MABU_ELEVATION and the SLOPE data layers. Your TABLE OF CONTENTS window should now look like the image at the top of the next page.

192

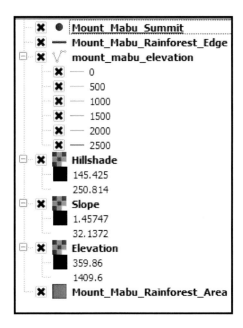

While the contents of your MAP window should look like this:

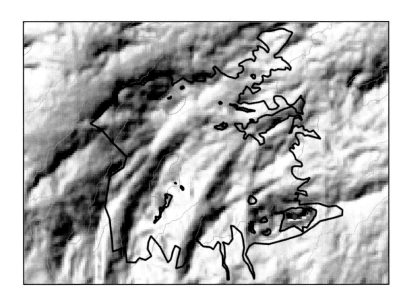

STEP 6: CREATE A RASTER DATA LAYER OF DISTANCE TO THE EDGE OF THE RAINFOREST:

So far, the raster data layers of environmental variables have all been derived from information on the elevation of the land around Mount Mabu. However, raster data layers can also be derived from many other types of environmental information. As a demonstration of this, in this step, a raster data layer will be created which measures the distance of each grid cell within the rainforest from the forest edge. This information can be used to identify areas where you are likely to find species which prefer to occur close to the edges of the forest, or near breaks in the forest canopy, and areas where you are likely to find species that only occur away from the forest edge. This raster data layer will be created in two stages. Firstly, a raster data layer will be created with a tool called R.GROW.DISTANCE which will calculate the distance that each grid cell is from the line marking the edge of the forest in the data layer MOUNT_MABU_RAINFOREST_EDGE. When this data layer is made, you will see that it calculates distance values for all cells in the raster data layer and not just the ones which are inside of the area of forest. In order to remove all the distance values from the cells outside the forest area, in the second stage, the polygon representing the area of forest in the data layer MOUNT_MABU_RAINFOREST_AREA is used to mask the distance raster data layer so that only the distance values inside the forest are retained. This then gives the final raster data layer where the values represent the distance of each forest grid cell to the nearest edge of the forest. To create this raster data layer, work through the following flow diagram:

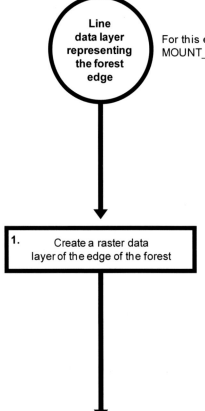

Line data layer representing the forest edge

For this exercise, the data layer representing the forest edge is MOUNT_MABU_RAINFOREST_EDGE.

Before you can create a raster data layer of distance to the forest edge, you first need to create a raster data layer of the forest edge itself. In the TOOLBOX window, select GRASS COMMANDS> VECTOR (V.*)> V.TO.RAST.ATTRIBUTE. In the V.TO.ATTRIBUTE window which opens, select MOUNT_MABU_RAINFOREST_EDGE as the INPUT VECTOR LAYER. For NAME OF COLUMN FOR 'ATT' PARAMETER, select ID. For GRASS REGION EXTENT, enter the following text:

207000,227000,8191000,8214000

1. Create a raster data layer of the edge of the forest

NOTE: There must be no spaces anywhere in this text otherwise this tool will not work. Next, enter 100 for GRASS REGION CELLSIZE. For RASTERIZED LAYER, leave the default setting of [SAVE TO TEMPORARY FILE]. Now click on RUN to run the tool. This will create a temporary raster data layer where each cell has a value indicating whether it contains the edge of the forest. If a cell does not contain the forest edge, it will be classified as 'no data' and will appear to be blank. This is okay.

2. Create a raster data layer of the distance to the edge of the forest

Next, in the TOOLBOX window, select GRASS COMMANDS> RASTER (R.*)> R.GROW.DISTANCE. In the tool window that opens, select RASTERIZED LAYER as the INPUT RASTER LAYER and EUCLIDEAN as the METRIC. Leave the GRASS REGION EXTENT with its default setting, but for GRASS REGION CELLSIZE enter 100. Next, in DISTANCE LAYER, type C:/GIS_FOR_BIOLOGISTS/ FOREST_EDGE.TIF and then click on the RUN button. **NOTE:** Do not worry if your GIS project appears to turn black when this tool finishes running. This is okay, and you will deal with this shortly.

Once the tool has finished running, right click on RASTERIZED LAYER in the TABLE OF CONTENTS window and select remove. Repeat this process to remove the OUTPUT VALUE data layer generated by the R.GROW.DISTANCE tool (this is the layer that makes your GIS project appear to be all black). Finally, right click on DISTANCE LAYER in the TABLE OF CONTENTS window and select RENAME. Type in the new name FOREST_EDGE and then press the ENTER key on your keyboard.

3. Mask your distance data layer with the polygon representing the area of forest using the CLIPPER tool

On the main menu bar, select RASTER> EXTRACTION> CLIPPER. This will open the CLIPPER tool window. In this window, select FOREST_EDGE as the INPUT FILE (RASTER), and then for OUTPUT FILE, enter C:/GIS_FOR_ BIOLOGISTS/FOREST_EDGE_2.TIF. Now, click on the box next to NO DATA VALUE to select it. Next, in the CLIPPING MODE section, select MASK LAYER and then select MOUNT_MABU_FOREST_AREA from the drop down menu in the section directly below it. Click OK to run the tool, and once it has finished running, click CLOSE to close the CLIPPER tool window. This will create a raster data layer where the values of the cells indicate their distance from the edge of the forest.

NOTE: All cells which fall exactly on the edge of the forest have a zero distance value and are treated as 'no data' cells. If you wanted to retain these cells in your raster data layer, you would need to process your distance raster data layer in a different way. This is by turning the forest area polygon into a raster data layer using the RASTERIZE tool (see step 7) and then using the RASTER CALCULATOR tool to combine these two data layers to give the required output.

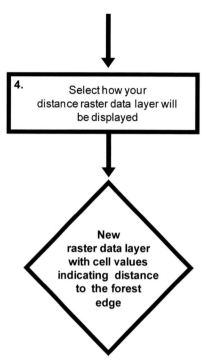

Right click on the name of the FOREST_EDGE_2 data layer in the TABLE OF CONTENTS window, and select PROPERTIES. In the LAYER PROPERTIES window, click on the STYLE tab on the left hand side. In the top left hand corner of the STYLE tab, select SINGLEBAND GRAY from the drop down menu next to RENDER TYPE. For COLOR GRADIENT, select BLACK TO WHITE and for CONTRAST ENHANCEMENT, select STRETCH TO MINMAX. Now, click OK to close the LAYER PROPERTIES window.

At the end of this step, change the position of the data layers called FOREST_EDGE and FOREST_EDGE_2 in your TABLE OF CONTENTS window by clicking on each one in turn and dragging it until it is between the MOUNT_MABU_ELEVATION and the HILLSHADE data layers. Your TABLE OF CONTENTS window should now look like this:

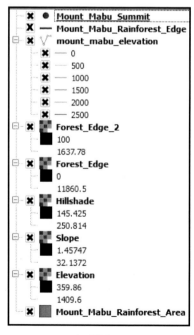

In the TABLE OF CONTENTS window, uncheck the boxes next to MOUNT_MABU_RAINFOREST_EDGE, MOUNT_MABU_RAIN FOREST_AREA and FOREST_EDGE_2 so that they are no longer displayed in the MAP window (but ensure that the FOREST_EDGE raster data layer is still set to display). At this stage, the contents of your MAP window should look like this:

Now, in the TABLE OF CONTENTS window, click on the box next to FOREST_EDGE_2 so that it is displayed again. Then, uncheck the boxes next to all the other raster data layers so that they are no longer displayed in the MAP window. This will allow you to see the effect of masking the FOREST_EDGE raster data layer to remove the cells which are outside of the rainforest area. The contents of your MAP window should look like this:

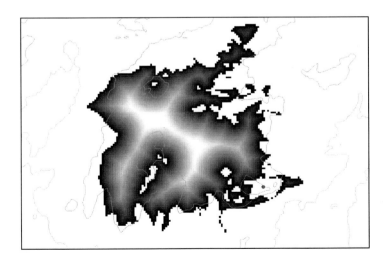

You can now remove the data layer called FOREST_EDGE from your GIS project as it is no longer needed. This is done by right-clicking on its name in the TABLE OF CONTENTS window and selecting REMOVE. The final FOREST_EDGE_2 raster data layer can be used to identify where species that prefer to occur well away from the edge of the forest are likely to occur, as well as where species which inhabit the forest edge are likely to be found.

STEP 7: CREATE A RASTER DATA LAYER REPRESENTING THE EXTENT OF THE RAINFOREST AREA ON MOUNT MABU IN 2013:

Up to this point in this exercise, you have been working with environmental data layers and using interpolation tools, such as the R.SURF.CONTOUR tool, to produce continuous surfaces when converting feature data layers to raster data layers. However, there is another way to create raster data layers from features, and this is to use what is known as a 'convert to grid' approach. In the 'interpolate grid' approach used for the environmental raster data layers, all the cells within the new raster data layer have values in them, with the tool interpolating values for any grid cells that do not contain a feature from the feature data layer (this was the reason that the first raster data layer of the distance to the forest edge had to be masked to remove the distance values from the cells which fell outside the area of rainforest). In contrast, in the 'convert to grid' approach, a grid cell only receives a value if it contains a feature from the feature data layer. All those cells that do no contain a feature are given a 'no data' value and this means that they are treated as empty cells in the final data layer.

To demonstrate the 'convert to grid' approach, you will create a raster data layer where cells which fall within the area covered by rainforest in 2013 have a value, and all other cells will be classified as 'no data'. This raster data layer will be made from the MOUNT_MABU_RAINFOREST_AREA polygon data layer which was originally created using the Google Earth user interface from an image taken in 2013, and which has been provided for this exercise. This means that this raster data layer will represent a baseline of forest coverage to which similar raster data layers from future images can be compared. Grid cells which have changed from being classified as containing rainforest to not containing rainforest between time periods can be used to identify not only whether deforestation has taken place, but in which grid cells it has occurred. Thus, this step creates a raster data layer which can be used to monitor the Mount Mabu rainforest area for potential changes which could indicate detrimental changes to this habitat, and so the organisms which occur there. In addition, by comparing these cells to the raster data layers of environmental variables, you would be able to tell whether changes to the rainforest coverage are concentrated in specific types of habitats, such as those close to the forest edge, those at lower altitudes or those in areas with gentle slopes which are easier to access and log.

Thus, the raster data layer of forest coverage which will be generated in this step could be used as part of a monitoring programme, with new raster data layers being created using the same methods whenever new images from this part of the world are available through the Google Earth user interface, or from any other suitable sources. This means that it is possible to monitor impacts of deforestation on Mount Mabu without having to necessarily visit the local area, allowing it to be conducted at relatively low cost.

To create this raster data layer of rainforest cover, work through the following flow diagram:

Polygon data layer which you wish to convert into a raster data layer

For this exercise, the polygon data layer which you will convert into a raster data layer using the 'convert to grid' approach will be MOUNT_MABU_RAINFOREST_AREA.

1. Create a raster data layer from the forest area data layer

On the main menu bar, select RASTER> CONVERSION> RASTERIZE (VECTOR TO RASTER). In the RASTERIZE tool window, select MOUNT_MABU_RAINFOREST_AREA for INPUT FILE (SHAPEFILE), and for ATTRIBUTE FIELD, select FID_FOREST. Now, for OUTPUT FILE FOR RASTERIZED VECTORS (RASTER), enter C:/GIS_FOR_BIOLOGISTS/FOREST_2013.TIF. Next, select RASTER RESOLUTION IN MAP UNITS PER PIXEL, and enter a value of 100 for both HORIZONTAL and VERTICAL before clicking on the EDIT button (it has a picture of a yellow pencil on it). Once you have entered editing mode, type a space after the existing code, and then enter the following new code into the box beside the EDIT button (leaving all the code that is already there unchanged):

-a_nodata 0 -te 207000 8191000 227000 8214000

NOTE: The term '-a_nodata 0' means that any cells that do not contain features from the feature data layer are classified as 'No Data' cells in the final raster data layer. The '-te' term defines the extent of the rater data layer that will be produced by the tool and sets the X Min, Y Min, X Max and Y Max values (in that order). Now click OK to run the tool. Once the tool has finished running, click CLOSE to close the RASTERIZE (VECTOR TO RASTER) TOOL window.

199

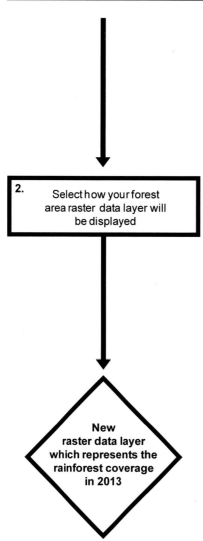

2. Select how your forest area raster data layer will be displayed

New raster data layer which represents the rainforest coverage in 2013

Right click on the name of the FOREST_2013 data layer in the TABLE OF CONTENTS window, and select PROPERTIES. In the LAYER PROPERTIES window, click on the STYLE tab on the left hand side. In the top left hand corner of the STYLE tab, select SINGLEBAND PSEUDOCOLOR from the drop down menu next to RENDER TYPE. Now, click on the plus sign (+) under COLOR INTERPOLATION to add a classification category to the table below it. In this table, enter 1 for value and FOREST AREA for LABEL. Finally, click on the box under COLOR and select a GREEN shade in the CHANGE COLOR window, before clicking OK to close this window and then clicking OK to close the LAYER PROPERTIES window.

At the end of this step, change the position of the data layer called FOREST_2013 in your TABLE OF CONTENTS window by clicking on it and dragging it until it is between the MOUNT_MABU_ELEVATION and the FOREST_EDGE_2 data layers. Your TABLE OF CONTENTS window should now look like the image at the top of the next page.

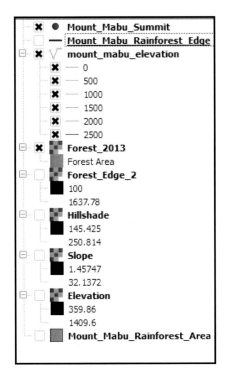

While the contents of your MAP window should look like this:

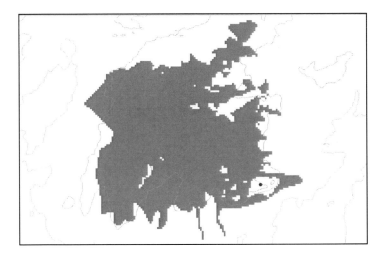

This is the end of exercise three. Now that you have completed it, you know how to create new raster data layers using two different approaches (the 'convert to grid' approach, used to create the raster data layer of the forest coverage, and the 'interpolate grid' approach used to create the elevation and distance to forest edge raster data layers). You also now know how to derive new data layers from existing raster data layers (such as the slope and the hillshade raster data layers, which were derived from the elevation raster data layer) and how to mask a raster data layer to remove unwanted values (as was done with the forest edge raster data layer). Together, these represent the main methods that biologists need to know to create new raster data layers in a GIS project. However, one further method (using a raster calculator tool) will be introduced in exercise six in chapter seventeen.

Chapter Fifteen

Exercise Four: How To Join Data From Different Layers Together

One of the main advantages of geographic information systems (GIS) over traditional, non-spatial databases is the ability to join information from different data layers together based on their spatial relationships. There are a number of different ways this can be done, and they include:

- **Spatial Joins:** A spatial join joins the information in the attribute tables of two different feature data layers together based on their spatial relationships. Spatial joins can be used either to link data together, or to summarize the information from one data layer based on features in another. For example, you can use a spatial join to count the number of points in a point data layer in each polygon of a polygon data layer. Spatial joins will be used in this exercise to count the number of dolphins recorded in each cell of a polygon grid data layer, and also to calculate the total amount of survey effort in each grid cell.

- **Unions:** While a spatial join joins information from attribute tables together, it does not affect the actual features in a feature data layer. In contrast, a union will result in changes to the features themselves. Specifically, a union will split any features which overlap into two or more new features based on how they overlap. For example, if you have a polygon representing a specific coastal study area, you could use a union to split the polygon into areas which fall on land and areas which fall in the sea using a polygon which represents areas of land. The resulting new data layer will consist of three polygons, one representing the section of the study area which is in the sea, one representing the section of the study area which falls on land, and a third which represents all other areas of land. This would allow sections of any polygon which fall on land to be removed, while retaining the parts of the polygons which fall in the sea.

- **Intersects:** An intersect is similar to a union. However, it splits the features of only one data layer by the features in another. For example, in the above union, if you did an intersect instead, you would be left with a single polygon representing only the section of the study area where it and the land overlap, and all other areas of both the land and the study area polygon would be removed. In this exercise, an intersect will be used to split lines representing survey effort at the locations where they cross the edges of polygon grid cells. This will then allow the total amount of survey effort in each grid cell to be calculated using a spatial join.

- **Extractions:** While spatial joins, unions and intersects can only work with feature data layers (i.e. those containing points, lines or polygons), extractions can be used to extract information from raster data layers. Specifically, it can be used to link information from raster data layers to the attribute tables of feature data layers. As such, they are very useful for linking environmental information contained in raster data layers (see exercise three) to point data representing locations which have been sampled, or where a specific species has been recorded.

- **Sampling:** Sampling also works with raster data layers. However, in contrast to extractions, they allow information from different raster data layers to be joined together. The main difference between sampling and extraction is that an extraction does not link the information to the attribute table of a data layer. Instead, it creates a new table which contains the joined information.

In this exercise, as an example of joining data from different data layers together, you will use an intersect and two spatial joins to bring together information from four different data layers to calculate the number of bottlenose dolphin recorded per kilometre of observations during surveys for cetaceans in northeast Scotland. These four data layers are:

1. **Polygon_Grid_North_Sea_WGS84.shp:** This is a polygon data layer representing 10km by 10km grid cells covering a study area that stretches from the northeast coastline of mainland Scotland to two groups of islands (the Orkney and the Shetland Islands) which lie further north. It is in a custom transverse mercator projection based on the WGS 84 datum called North Sea, which has been created specifically for this study. The fact that it is in this custom projection/coordinate system is reflected in the suffix (North_Sea_WGS84) added to the data layer's name. It is this custom projection/coordinate system which will be used for this exercise.

2. **Survey_Tracks_North_Sea_WGS84.shp:** This is a line data layer which represents the route of real surveys taken between the city of Aberdeen on mainland Scotland to the two island groups which lie further north. This data layer is in a custom transverse mercator projection based on the WGS 84 datum called North Sea, which has been created specifically for this study. The fact that it is in this custom projection/coordinate system is reflected in the suffix (North_Sea_WGS84) added to the data layer's name.

3. **Bottlenose_Dolphin_North_Sea_WGS84.shp:** This is a point data layer which represents locations where bottlenose dolphins were recorded during these surveys. This data layer is in a custom transverse mercator projection based on the WGS 84 datum called North Sea, which has been created specifically for this study. The fact that it is in this custom projection/ coordinate system is reflected in the suffix (North_Sea_WGS84) added to the data layer's name.

4. **Land_North_Sea_WGS84.shp:** This is a polygon data layer which represents areas of land in the study area. This is provided for information purposes only and will not be used during the actual processing. This data layer is in a custom transverse mercator projection based on the WGS 84 datum called North Sea, which has been created specifically for this study. The fact that it is in this custom projection/coordinate system is reflected in the suffix (North_Sea_WGS84) added to the data layer's name.

The instructions for calculating the abundance of bottlenose dolphin per unit effort from survey data in ArcGIS 10.3 can be found on page 206, while those for doing the same thing in QGIS 2.8.3 can be found on page 224. However, if you have not already done so, before you start this exercise, you will first need to create a new folder on your C: drive called GIS_FOR_BIOLOGISTS. To do this on a computer with a Windows operating system, open Windows Explorer and navigate to your C:\ drive (this may be called Windows C:). To create a new folder on this drive, right click on the window displaying the contents of your C:\ drive and select NEW> FOLDER. This will create a new folder. Now call this folder GIS_FOR_BIOLOGISTS by typing this into the folder name to replace what it is currently called (which will most likely be NEW FOLDER). This folder, which has the address C:\GIS_FOR_BIOLOGISTS, will be used to store all files and data for the exercises in this book.

Next, you need to download the source files for the four data layers outlined above from *www.gisinecology.com/GFB.htm#2*. Once you have downloaded the compressed folder containing the files, make sure that you then copy all the files it contains into the folder C:\GIS_FOR_BIOLOGISTS. **NOTE:** If you have already downloaded the compressed file from *GISinEcology.com* for any previous exercise, you will find that you have already downloaded these files, and you do not need to do this again.

Instructions For ArcGIS 10.3 Users:

Once you have all the required files downloaded into the correct folder on your computer, and you understand what is contained within each file, you can move on to creating your GIS project. The starting point for this is a blank GIS project. To create a blank GIS project, first start the ArcGIS software by opening the ArcMap module. When it opens, you will be presented with a window which has the heading ARCMAP – GETTING STARTED. In this window, you can either select an existing GIS project to work on, or create a new one. To create a new blank GIS project, click on NEW MAPS in the directory tree on the left hand side and then select BLANK MAP in the right hand section of the window. Now, click OK at the bottom of this window. This will open a new blank GIS project. (**NOTE:** If this window does not appear, you can start a new project by clicking on FILE on the main menu bar, and selecting NEW. When the NEW DOCUMENT window opens, select NEW MAPS and then BLANK MAP in the order outlined above.) Once you have opened your new GIS project, the first thing you need to do is save it under a new, and meaningful, name. To do this, click on FILE on the main menu bar, and select SAVE AS. Save it as EXERCISE_FOUR in the folder C:\GIS_FOR_BIOLOGISTS. Remember to save your project again at the end of each and every step (you will not be reminded to do this).

STEP 1: SET THE PROJECTION/COORDINATE SYSTEM FOR YOUR DATA FRAME:

Whenever you start a new GIS project, the first thing you should do, even before you add any data layers, is select an appropriate projection/coordinate system and then set the data frame (i.e. the map you are going to add your data layers to) to use this projection/coordinate system. This ensures that you do not accidently end up using one which is inappropriate. In addition, most problems that beginners have with GIS are caused by a failure to use an appropriate projection/coordinate system or a conflict between the projection/coordinate system used for data layers and the one being used for the data frame itself. For this exercise, the projection/coordinate system which will be used is a custom transverse mercator called North Sea that is centred on latitude 56.5°N and longitude 1.0°W (or latitude 56.5° and longitude -1.0° in decimal degrees) based on the WGS 1984 datum. This is the projection/coordinate system which you will set your data frame to in step one, and this can be done by working through the following flow diagram:

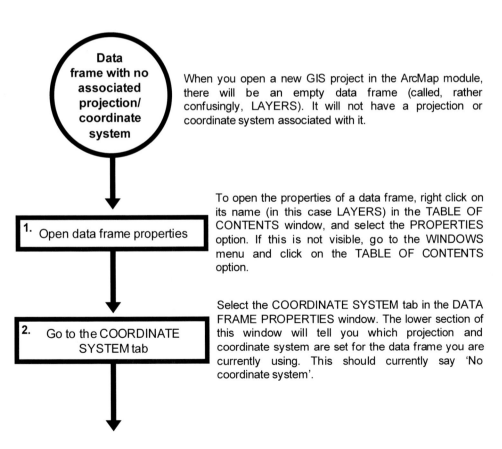

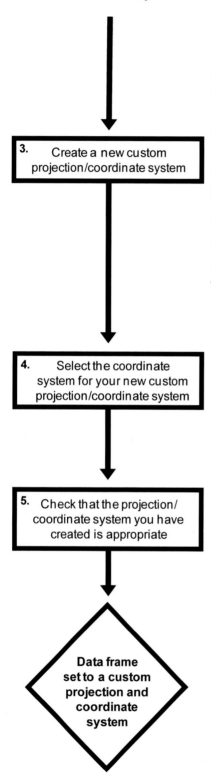

3. Create a new custom projection/coordinate system

4. Select the coordinate system for your new custom projection/coordinate system

5. Check that the projection/coordinate system you have created is appropriate

Data frame set to a custom projection and coordinate system

To create a new custom projection/coordinate system, click on the ADD COORDINATE SYSTEM button that can be found towards the top right hand corner of the COORDINATE SYSTEM tab (it has a picture of a globe on it), and select NEW> PROJECTED COORDINATE SYSTEM. This will open the NEW PROJECTED COORDINATE SYSTEM window. In the upper NAME section of the window, type: North Sea. **NOTE:** This must be entered exactly as written, with capital letters at the start of each word, and lower case letters for the rest. In the PROJECTION portion of the window, select the name of the appropriate type of coordinate system from the drop down menu (in the lower NAME section). For this exercise, select TRANSVERSE MERCATOR. Next, type in the values you wish to use for the parameters. For LATITUDE_ OF_ORIGIN enter 56.5, and for CENTRAL_ MERIDIAN enter -1.0. Leave all other sections of the window with their default settings.

In the GEOGRAPHIC COORDINATE SYSTEM section of the PROJECTED COORDINATE SYSTEM window, by default it should say NAME: GCS_WGS_1984. If it doesn't, click on the CHANGE button and type WGS 1984 into the SEARCH box in the window that appears and press the ENTER key on your keyboard. Select WORLD> WGS 1984, and click the OK button. Now click the OK button in the NEW PROJECTED COORDINATE SYSTEM window. Finally, click OK in the DATA FRAME PROPERTIES window.

Once you have created a custom projection/ coordinate system, you need to check that it is appropriate. This involves examining how data layers look in it. For this exercise, this will be done in the next step by adding a polygon data layer of land and checking that it looks the right shape.

208

To check that you have done this step properly, right click on the name of your data frame (LAYERS) in the TABLE OF CONTENTS window and select PROPERTIES. Click on the COORDINATE SYSTEM tab and make sure that the contents of the CURRENT COORDINATE SYSTEM section of the window has the following text at the top of it:

> North Sea
> Authority: Custom
>
> Projection: Transverse_Mercator
> False_Easting: 0.0
> False_Northing: 0.0
> Central_Meridian: -1.0
> Scale_Factor: 1.0
> Latitude_Of_Origin: 56.5
> Linear Unit: Meter (1.0)

If it does not, you will need to repeat this step to ensure that you have assigned the correct projection/coordinate system to your data frame.

Now click OK to close the DATA FRAME PROPERTIES window.

STEP 2: ADD THE REQUIRED EXISTING DATA LAYERS TO YOUR GIS PROJECT:

Once the projection/coordinate system has been set for your data frame, you need to add the data layers which you will use for this exercise. These are POLYGON_GRID_NORTH_SEA_WGS84.SHP, SURVEY_TRACKS_NORTH_SEA_WGS84.SHP, BOTTLENOSE_DOLPHIN_NORTH_SEA_WGS84.SHP and LAND_NORTH_SEA_WGS84.SHP. **NOTE:** The data layer LAND_NORTH_SEA_WGS84 will not be directly used in this exercise, but is added to provide additional information for display purposes. To add these data layers to your GIS project, work through the flow diagram which starts on the next page.

Existing data layers which you wish to add to your GIS project

The data layers you wish to add are POLYGON_ GRID_NORTH_SEA_WGS84.SHP, SURVEY_TRACKS _NORTH_SEA_WGS84.SHP, BOTTLENOSE_DOLPHIN _NORTH_SEA_WGS84.SHP and LAND_NORTH_SEA _WGS84.SHP.

1. Make sure that the data frame to which you wish to add the data layer is active

Right click on the name of the data frame (in this case LAYERS) which you wish to add the data to in the TABLE OF CONTENTS window, and select ACTIVATE (it will appear as if nothing has happened, but the data frame will now be active).

2. Open the ADD DATA window

Right click on the name of the data frame (LAYERS) in the TABLE OF CONTENTS window, and select ADD DATA. In the ADD DATA window, browse to the location of your data layer (C:\GIS_FOR_BIOLOGISTS) and select the data layer called POLYGON_GRID_NORTH_SEA_ WGS84.SHP. Now click ADD. A polygon data layer will appear in your MAP window and the data layer named POLYGON_GRID_NORTH_SEA_ WGS84 will have been added to the TABLE OF CONTENTS window.

3. Check the projection/ coordinate system for the newly added data layer

Whenever you add a data layer to a GIS project, you should always check that it has a projection/ coordinate system assigned to it, and look at what this projection/coordinate system is. This is so that you know whether you will need to assign a projection/coordinate system to it, or transform it into a different projection/coordinate system before you can use it in your GIS project. To check the projection/coordinate system of your newly added data layer, right click on its name in the TABLE OF CONTENTS window and select PROPERTIES. In the LAYER PROPERTIES window which opens, click on the SOURCE tab and check that the projection/coordinate system listed under DATA SOURCE is NORTH SEA (the same as the data frame). Now click OK to close the LAYER PROPERTIES window. If the data layer had not been in the same projection/ coordinate system as the data frame, you would have to have used the PROJECT tool to transform it into the correct projection/coordinate system (see exercise five for more details).

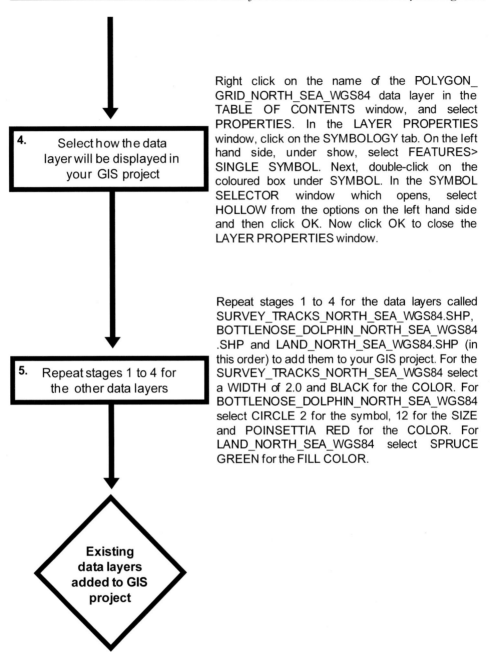

4. Select how the data layer will be displayed in your GIS project

Right click on the name of the POLYGON_ GRID_NORTH_SEA_WGS84 data layer in the TABLE OF CONTENTS window, and select PROPERTIES. In the LAYER PROPERTIES window, click on the SYMBOLOGY tab. On the left hand side, under show, select FEATURES> SINGLE SYMBOL. Next, double-click on the coloured box under SYMBOL. In the SYMBOL SELECTOR window which opens, select HOLLOW from the options on the left hand side and then click OK. Now click OK to close the LAYER PROPERTIES window.

5. Repeat stages 1 to 4 for the other data layers

Repeat stages 1 to 4 for the data layers called SURVEY_TRACKS_NORTH_SEA_WGS84.SHP, BOTTLENOSE_DOLPHIN_NORTH_SEA_WGS84 .SHP and LAND_NORTH_SEA_WGS84.SHP (in this order) to add them to your GIS project. For the SURVEY_TRACKS_NORTH_SEA_WGS84 select a WIDTH of 2.0 and BLACK for the COLOR. For BOTTLENOSE_DOLPHIN_NORTH_SEA_WGS84 select CIRCLE 2 for the symbol, 12 for the SIZE and POINSETTIA RED for the COLOR. For LAND_NORTH_SEA_WGS84 select SPRUCE GREEN for the FILL COLOR.

Existing data layers added to GIS project

At this point, your TABLE OF CONTENTS window should look like the figure at the top of the next page.

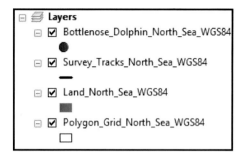

Now, right click on POLYGON_GRID_NORTH_SEA_WGS84 in the TABLE OF CONTENTS window and select ZOOM TO LAYER. The contents of your MAP window should now look like this:

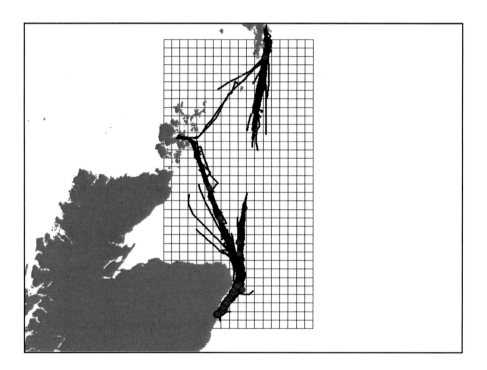

If the contents of your MAP window do not look like this, look in the TABLE OF CONTENTS window and check that you have all the required data layers added to your GIS project and that they are in the right order. If any data layers are missing, repeat step 2 until all the required data layers are present.

STEP 3: CALCULATE THE LENGTH OF SURVEY TRACKS IN EACH POLYGON GRID CELL:

The next step in this exercise is to divide the survey tracks into sections which fall into each grid cell and then calculate the lengths of all these sections. This is done using the INTERSECT tool as outlined below. Once this has been calculated, this information can be joined to the attribute table of the polygon grid data layer using the SPATIAL JOIN tool. To carry out the required intersect and spatial join, work through the flow diagram below. **NOTE:** These instructions assume that you have the EDITOR toolbar set to display in the OPTIONAL TOOL BARS area of the ArcMap user interface. If this is not the case, click on CUSTOMIZE on the main menu bar, select TOOLBARS and make sure that there is a tick next to the EDITOR toolbar option.

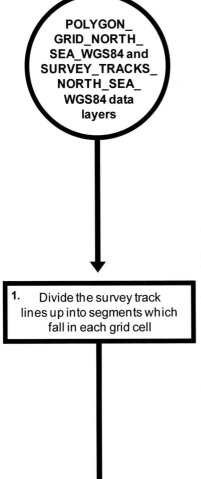

POLYGON_ GRID_NORTH_ SEA_WGS84 and SURVEY_TRACKS_ NORTH_SEA_ WGS84 data layers

1. Divide the survey track lines up into segments which fall in each grid cell

In the TOOLBOX window, select ANALYSIS TOOLS> OVERLAY> INTERSECT. This will open the INTERSECT tool window. In the INTERSECT tool window, select SURVEY_TRACKS_NORTH_SEA_ WGS84 from the drop down menu in the INPUT FEATURES section. Once selected, it will appear in the FEATURES section immediately below the INPUT FEATURES section. Repeat this for POLYGON_ GRID_NORTH_SEA_WGS84. Next, type C:\GIS_ FOR_BIOLOGISTS\SURVEY_TRACKS_INTERSECT in the OUTPUT FEATURE CLASS section of the window. In the JOIN ATTRIBUTES section, select ALL from the drop down menu. In the OUTPUT TYPE section, select INPUT from the drop down menu. Finally, click OK to run the INTERSECT tool window.

2. Add field called LENGTH and calculate the length of each newly created segment

3. Remove any new lines which have a zero length

Right click on the name SURVEY_ TRACKS_INTERSECT in the TABLE OF CONTENTS window, and select OPEN ATTRIBUTE TABLE. Click on the OPTIONS button at the top left corner of the TABLE window and select ADD FIELD. Name the field 'Length', and select SHORT INTEGER for the type. Type in 16 for PRECISION. Now click OK. Next, right-click on the field name LENGTH in the TABLE window and select CALCULATE GEOMETRY. If a window appears warning you that you are editing a data layer outside of an edit session, click YES, and carry on. If this window does not appear, this is OK. In the CALCULATE GEOMETRY window, for PROPERTY, select LENGTH. For COORDINATE SYSTEM select USE COORDINATE SYSTEM OF THE DATA SOURCE. In UNITS, select METRES. Once you have done this, click OK to run the CALCULATE GEOMETRY tool.

You now want to remove any line features from the data layer which have a zero length. To do this, go to SELECTION on the main menu bar and select SELECT BY ATTRIBUTES. In the SELECT BY ATTRIBUTES window, select SURVEY_TRACKS_ INTERSECT for LAYER and CREATE A NEW SELECTION for METHOD. Double click on the field name LENGTH to add it to the lower window. Now click on the equals (=) sign to add it to the lower window before typing in a space followed by the number zero (0). This will result in the expression "LENGTH" = 0 appearing in the lower window. Now click OK.

On the EDITOR tool bar, click the EDITOR button and select START EDITING. The EDITOR window should now appear at the right hand side of the MAP window (it may be called CREATE FEATURES). If the EDITOR window does not appear automatically, you can open it by clicking on the EDITOR button and selecting EDITING WINDOWS> CREATE FEATURES. If the name of the data layer SURVEY_ TRACKS_INTERSECT appears in this window, select it by clicking on it once. If it doesn't, click on the ORGANIZE TEMPLATES button at the top of the EDITOR window (it should be second from the left). Click on the NEW TEMPLATE button at the top of the ORGANIZE FEATURE TEMPLATE window and select the data layer SURVEY_TRACKS_ INTERSECT, then click on the FINISH button. This data layer will now appear in the EDITOR window and you will be able to select it. Next, click on EDIT on the main menu bar and select DELETE. Click on the EDITOR button on the EDITOR tool bar again and select STOP EDITING. When the SAVE window opens click on YES to save the edits.

214

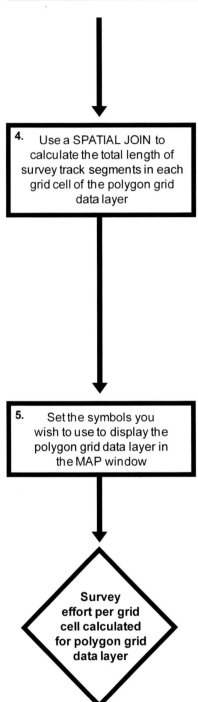

4. Use a SPATIAL JOIN to calculate the total length of survey track segments in each grid cell of the polygon grid data layer

5. Set the symbols you wish to use to display the polygon grid data layer in the MAP window

Survey effort per grid cell calculated for polygon grid data layer

In the TOOLBOX window, select ANALYSIS TOOLS> OVERLAY> SPATIAL JOIN. In the SPATIAL JOIN window, select the POLYGON_GRID_NORTH_ SEA_WGS84 in the TARGET FEATURES section, and select SURVEY_TRACKS_INTERSECT in the JOIN FEATURES section. Next, in the OUTPUT FEATURE CLASS section, type C:\GIS_FOR_ BIOLOGISTS\POLYGON_GRID_NORTH_SEA_SUR VEY_EFFORT.

In the JOIN OPERATION (OPTIONAL) section of the window, select JOIN_ONE_TO_ONE. In the FIELD MAP OF JOIN FEATURES (OPTIONAL) section, right click on the field called LENGTH and select MERGE RULE> SUM. Remove all fields except FID_ POLYGO, LENGTH and CELL_ID_NO. This is done by right-clicking on a field's name and selecting DELETE. Scroll down until you see the section called MATCH OPTION (OPTIONAL) and select CONTAINS from its drop down menu. Finally, click OK.

Right click on the name of your newly created polygon grid data layer (POLYGON_GRID_NORTH_SEA_ SURVEY_EFFORT) in the TABLE OF CONTENTS window and select PROPERTIES. Next, click on the SYMBOLOGY tab of the LAYER PROPERTIES window. In the left hand portion of the LAYER PROPERTIES window, select QUANTITIES> GRADUATED COLORS. From the drop down menu beside where is says COLOR RAMP, select the option that goes from white to black (you may have to scroll down to find it). Next to VALUE, select LENGTH. Under RANGE, click on the top line and type in 0, then click on the next line and type in 10000. Type in 50000 for the next, 100000 for the next and 5000000 for the last one. Finally, click OK to close the LAYER PROPERTIES window.

Once you have finished working through this flow diagram, remove the data layers SURVEY_TRACKS_INTERSECT and POLYGON_GRID_NORTH _SEA_WGS84 from your GIS project by right-clicking on their names in the

215

TABLE OF CONTENTS window and selecting REMOVE. Finally, change the order of the data layers by clicking on their names in the TABLE OF CONTENTS window and, while holding the mouse button down, dragging them upwards or downwards to their desired position so that your TABLE OF CONTENTS window looks like this:

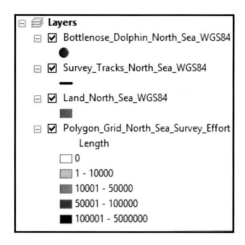

The ATTRIBUTE TABLE for the data layer POLYGON_GRID_ NORTH_SEA_SURVEY_EFFORT should now either look like the figure below or, more likely, the one at the top of the next page. (**NOTE:** The OBJECT ID field may be called the FID field.)

OBJECTID *	Shape *	Join_Count	TARGET_FID	FID_polygo	CELL_ID_NO	LENGTH	Shape_Length	Shape_Area
1	Polygon	0	0	0	1	0	40000	100000000
2	Polygon	0	1	1	2	0	40000	100000000
3	Polygon	0	2	2	3	0	40000	100000000
4	Polygon	0	3	3	4	0	40000	100000000
5	Polygon	0	4	4	5	0	40000	100000000
6	Polygon	0	5	5	6	0	40000	100000000
7	Polygon	0	6	6	7	0	40000	100000000
8	Polygon	0	7	7	8	0	40000	100000000
9	Polygon	0	8	8	9	0	40000	100000000
10	Polygon	0	9	9	10	0	40000	100000000
11	Polygon	0	10	10	11	0	47892.904314	98230232.244097
12	Polygon	0	11	11	12	0	89331.215494	37940633.23534
13	Polygon	446	12	12	13	1727287	44441.675602	98218513.217431
14	Polygon	0	13	13	14	0	40000	100000000
15	Polygon	0	14	14	15	0	40000	100000000
16	Polygon	0	15	15	16	0	40000	100000000
17	Polygon	0	16	16	17	0	40000	100000000
18	Polygon	0	17	17	18	0	40000	100000000
19	Polygon	0	18	18	19	0	40000	100000000
20	Polygon	0	19	19	20	0	40000	100000000

OBJECTID *	Shape *	Join_Count	TARGET_FID	FID_polygo	CELL_ID_NO	LENGTH
1	Polygon	0	0	0	1	0
2	Polygon	0	1	1	2	0
3	Polygon	0	2	2	3	0
4	Polygon	0	3	3	4	0
5	Polygon	0	4	4	5	0
6	Polygon	0	5	5	6	0
7	Polygon	0	6	6	7	0
8	Polygon	0	7	7	8	0
9	Polygon	0	8	8	9	0
10	Polygon	0	9	9	10	0
11	Polygon	0	10	10	11	0
12	Polygon	0	11	11	12	0
13	Polygon	446	12	12	13	1727287
14	Polygon	0	13	13	14	0
15	Polygon	0	14	14	15	0
16	Polygon	0	15	15	16	0
17	Polygon	0	16	16	17	0
18	Polygon	0	17	17	18	0
19	Polygon	0	18	18	19	0
20	Polygon	0	19	19	20	0

Finally, click on the box to the left of the data layer called SURVEY_TRACKS _NORTH_SEA_WSGS84 in the TABLE OF CONTENTS window so that the tick in it disappears and this data layer is no longer displayed in the MAP window.

The contents of your MAP window should look like this:

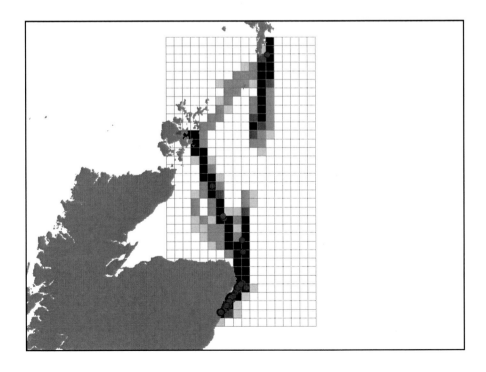

STEP 4: CALCULATE THE NUMBER OF BOTTLENOSE DOLPHINS RECORDED IN EACH GRID CELL:

The next data which must be added to the attribute table of the polygon grid data layer is information on the number of bottlenose dolphins recorded in each grid cell. This is also added using a SPATIAL JOIN to calculate the total number of dolphins for all sightings which fall within each grid cell. To do this, work through the following flow diagram:

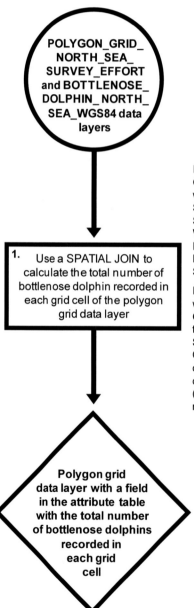

In the TOOLBOX window, select ANALYSIS TOOLS> OVERLAY> SPATIAL JOIN. In the SPATIAL JOIN window, select the POLYGON_GRID_NORTH_SEA_ SURVEY_EFFORT in the TARGET FEATURES section. Select BOTTLENOSE_DOLPHIN_NORTH_SEA_ WGS84 in the JOIN FEATURES section. In the OUTPUT FEATURE CLASS section, type C:\GIS_FOR_ BIOLOGISTS\POLYGON_GRID_NORTH_SEA_WITH_ SIGHTINGS.

In the JOIN OPERATION (OPTIONAL) section of the window select JOIN_ONE_TO_ONE. In the FIELD MAP OF JOIN FEATURES (OPTIONAL) section, right click on the field called NUMBER and select MERGE RULE> SUM. Remove all fields except FID_POLYGO, LENGTH, CELL_ID_NO and NUMBER. This is done by right-clicking on a field's name and selecting DELETE. Scroll down until you see the section called MATCH OPTION (OPTIONAL) and select CONTAINS from its drop down menu. Finally, click OK.

At the end of this step, remove the data layer called POLYGON_ GRID_NORTH_SEA_SURVEY_EFFORT by right-clicking on its name in the TABLE OF CONTENTS window and selecting REMOVE. Finally, open the TABLE window for the data layer POLYGON_GRID_ NORTH_SEA_ WITH_SIGHTINGS by right-clicking on its name in the TABLE OF CONTENTS window and selecting OPEN ATTRIBUTE TABLE. The attribute table should look like this:

FID	Shape *	Join_Count	TARGET_FID	FID_polygo	CELL_ID_NO	LENGTH	Number
0	Polygon	0	1	0	1	0	0
1	Polygon	0	2	1	2	0	0
2	Polygon	0	3	2	3	0	0
3	Polygon	0	4	3	4	0	0
4	Polygon	0	5	4	5	0	0
5	Polygon	0	6	5	6	0	0
6	Polygon	0	7	6	7	0	0
7	Polygon	0	8	7	8	0	0
8	Polygon	0	9	8	9	0	0
9	Polygon	0	10	9	10	0	0
10	Polygon	0	11	10	11	0	0
11	Polygon	0	12	11	12	0	0
12	Polygon	0	13	12	13	1727287	0
13	Polygon	0	14	13	14	0	0
14	Polygon	0	15	14	15	0	0
15	Polygon	0	16	15	16	0	0
16	Polygon	0	17	16	17	0	0
17	Polygon	0	18	17	18	0	0
18	Polygon	0	19	18	19	0	0
19	Polygon	0	20	19	20	0	0
20	Polygon	0	21	20	21	0	0

Now close the attribute table.

STEP 5: CALCULATE THE ABUNDANCE PER UNIT EFFORT IN EACH GRID CELL:

You now have all the information in the attribute table of the polygon grid data layer which you need to calculate the abundance of bottlenose dolphins per kilometre of survey effort in each grid cell. This will be calculated by adding a new field and using the FIELD CALCULATOR tool to divide the number of bottlenose dolphins recorded in each grid cell by the amount of survey effort. This can be done by working through the flow diagram which starts on the next page.

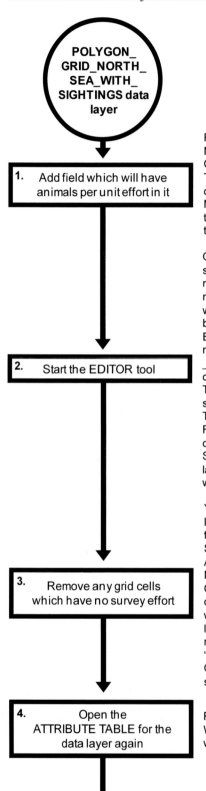

POLYGON_ GRID_NORTH_ SEA_WITH_ SIGHTINGS data layer

1. Add field which will have animals per unit effort in it

2. Start the EDITOR tool

3. Remove any grid cells which have no survey effort

4. Open the ATTRIBUTE TABLE for the data layer again

Right click on the name POLYGON_GRID_ NORTH_SEA_WITH_SIGHTINGS in the TABLE OF CONTENTS window, and select OPEN ATTRIBUTE TABLE. Click on the OPTIONS button at the top left corner of the TABLE window and select ADD FIELD. Name the field 'Animals_Km' and select DOUBLE for the type. Type in 16 for PRECISION and 6 for the scale, then click OK before closing the TABLE window.

On the EDITOR toolbar, click the EDITOR button and select START EDITING. The EDITOR window should now appear at the right hand side of the MAP window. (it may be called CREATE FEATURES). If the EDITOR window does not appear automatically, you can open it by clicking on the EDITOR button and selecting EDITING WINDOWS> CREATE FEATURES. If the name of the data layer POLYGON_GRID_NORTH_SEA _WITH_SIGHTINGS appears in this window, select it by clicking on it once. If it doesn't, click on the ORGANIZE TEMPLATES button at the top of the EDITOR window (it should be second from the left). Click on the NEW TEMPLATE button at the top of the ORGANIZE FEATURE TEMPLATE window and select the data layer called POLYGON_GRID_NORTH_SEA_WITH_ SIGHTINGS, then click on the FINISH button. This data layer will now appear in the EDITOR window and you will be able to select it.

You now want to remove any grid cells from the data layer which have no survey effort in them. To do this, go to SELECTION on the main menu bar and select SELECT BY ATTRIBUTES. In the SELECT BY ATTRIBUTES window, select POLYGON_GRID_ NORTH_SEA_WITH_SIGHTINGS for LAYER and CREATE A NEW SELECTION for METHOD. Double click on the field name LENGTH to add it to the lower window. Now click on the equals (=) sign to add it to the lower window before typing in a space followed by the number zero (0). This will result in the expression "LENGTH" = 0 appearing in the lower window. Now click OK. Next, click on EDIT on the main menu bar and select DELETE.

Right click on POLYGON_GRID_NORTH_SEA_ WITH_SIGHTINGS in the TABLE OF CONTENTS window and select OPEN ATTRIBUTE TABLE.

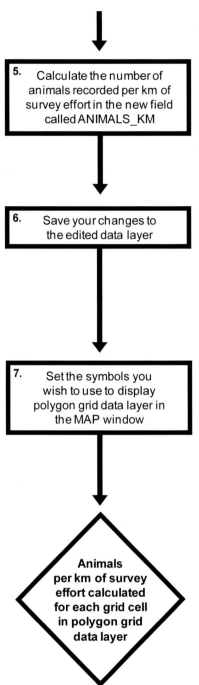

5. Calculate the number of animals recorded per km of survey effort in the new field called ANIMALS_KM

6. Save your changes to the edited data layer

7. Set the symbols you wish to use to display polygon grid data layer in the MAP window

Animals per km of survey effort calculated for each grid cell in polygon grid data layer

In the TABLE window, right click on the field ANIMALS_KM and select FIELD CALCULATOR. This will open the FIELD CALCULATOR window. In the lower part of this window enter the expression: [NUMBER] / ([LENGTH] / 1000). The expression needs to be entered exactly like this (spaces and all), and make sure that you have the brackets in exactly the right place, or you will not get the correct result). Now click OK. **NOTE:** The '/ 1000' term is included to convert the survey effort data from metres (the units selected when applying the CALCULATE GEOMETRY tool) to kilometres.

Once you have finished calculating the number of animals per km of survey effort, you need to save the edits to your data layer. To do this, first close the attribute table. Next, on the EDITOR toolbar, click on EDITOR and select SAVE EDITS. Next, in the EDITOR toolbar, click on the EDITOR button and select STOP EDITING.

Right click on the name of your polygon grid data layer (POLYGON_GRID_NORTH_SEA_WITH_SIGHTINGS) in the TABLE OF CONTENTS window and select PROPERTIES. Next, click on the SYMBOLOGY tab of the LAYER PROPERTIES window. In the left hand portion of the LAYER PROPERTIES window, select QUANTITIES> GRADUATED COLOURS. Beside VALUE, select ANIMALS_KM. Under RANGE, click on the top line and type in 0, then click on the next line and type in 0.001. Type in 0.005 for the next, 0.01 for the next and 1.00 for the last one. Now click on the coloured box next to where it says 0 under RANGE and select white for the fill colour (it may already be set to this colour) in the SYMBOL SELECTOR window. Finally, click OK to close the SYMBOL SELECTOR window and then click OK to close the LAYER PROPERTIES window.

Once you have calculated the recorded abundance of bottlenose dolphin per kilometre of survey effort per grid cell, you need to turn off all the other data layers with the exception of LAND_NORTH_SEA_WS84 and POLYGON_ GRID_NORTH_SEA_WITH_SIGHTINGS.

Right click on the name of the data layer POLYGON_GRID_NORTH_SEA_WITH_SIGHTINGS in the TABLE OF CONTENTS window and select OPEN ATTRIBUTE TABLE. The ATTRIBUTE TABLE for POLYGON_GRID_NORTH_SEA_WITH_SIGHTINGS should look like this:

FID	Shape *	Join_Count	TARGET_FID	FID_polygo	CELL_ID_NO	LENGTH	Number	Animals_km
0	Polygon	0	13	12	13	1727287	0	0
1	Polygon	0	30	29	30	2852	0	0
2	Polygon	0	31	30	31	1748827	0	0
3	Polygon	0	48	47	48	96869	0	0
4	Polygon	1	49	48	49	1626920	25	0.015366
5	Polygon	0	64	63	64	1096	0	0
6	Polygon	0	65	64	65	32774	0	0
7	Polygon	0	66	65	66	614236	0	0
8	Polygon	0	67	66	67	1038366	0	0
9	Polygon	0	81	80	81	12738	0	0
10	Polygon	0	82	81	82	25747	0	0
11	Polygon	0	83	82	83	15115	0	0
12	Polygon	0	84	83	84	1318525	0	0
13	Polygon	0	85	84	85	184766	0	0
14	Polygon	0	97	96	97	5021	0	0
15	Polygon	0	98	97	98	11561	0	0
16	Polygon	0	99	98	99	23209	0	0
17	Polygon	0	100	99	100	12789	0	0
18	Polygon	0	101	100	101	2457	0	0
19	Polygon	0	102	101	102	1333776	0	0
20	Polygon	0	103	102	103	58460	0	0

Now close the TABLE window. At this stage, your TABLE OF CONTENTS window should look like this:

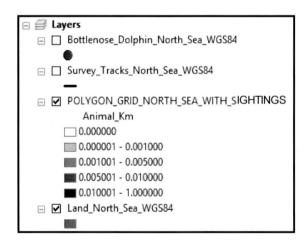

And the contents of your MAP window should look like the image below (**NOTE:** The exact colour of each grid cell may vary from this image due to the printing process, but the relative shading of each one should be the same).

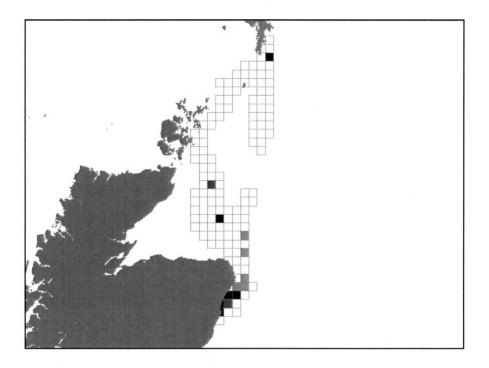

This is the end of exercise four. In this exercise, you have learned how to join information from different data layers together in two different ways. In the first, using an INTERSECT allowed you to use the information in one data layer to split the features in a second data layer (in this case, cut the lines representing survey effort at the points where they cross from one grid cell in the polygon grid data layer into another). In the second way, you used a SPATIAL JOIN, which allowed you to calculate both the total amount of survey effort in each cell of the polygon grid, and also the total number of bottlenose dolphin seen in each grid cell. The other ways of joining information from different data layers together, such as UNION, and EXTRACTION, work in very similar ways, but provide slightly different outputs. As a result, it is important to select the right way for joining information from different data layers together, depending on the exact outcome you wish to achieve.

Instructions for QGIS 2.8.3 Users:

Once you have all the required files downloaded into the correct folder on your computer, and you understand what is contained within each file, you can move on to creating your GIS project for this exercise. The starting point for this is a blank GIS project. To create a blank GIS project, first open QGIS. Once it is open, click on the PROJECT menu and select SAVE AS. In the window which opens, save your GIS project as EXERCISE_FOUR in the folder C:\GIS_FOR_ BIOLOGISTS. Remember to save your project again at the end of each and every step (you will not be reminded to do this).

STEP 1: SET THE PROJECTION/COORDINATE SYSTEM FOR YOUR DATA FRAME:

Whenever you start a new GIS project, the first thing you should do, even before you add any data layers, is select an appropriate projection and/coordinate system, and then set the data frame (i.e. the map you are going to add your data layers to) to use this projection/coordinate system. This ensures that you do not accidently end up using one which is inappropriate. In addition, most problems that beginners have with GIS are caused by a failure to use an appropriate projection/coordinate system or a conflict between the projection/coordinate system used for data layers and the one being used for the data frame itself. For this exercise, the projection/coordinate system that will be used is the custom transverse mercator called North Sea which is centred on latitude 56.5°N and longitude 1.0°W (or latitude 56.5° and longitude -1.0° in decimal degrees) based on the WGS 1984 datum introduced in exercise one. This is the projection/coordinate system which you will set your data frame to in step one, and this can be done by working through the flow diagram which starts on the next page. **NOTE:** In QGIS, the projection/coordinate systems are set using a short piece of code known as a Proj.4 string (see *http://github.com/Oseo/proj.4/wiki* for more details). It is very easy to enter these Proj.4 string codes incorrectly. As a result, the Proj.4 string code for the custom transverse projection/coordinate system which will be used in this exercise has been included in a file in the compressed data folder for this book. It is called North_Sea_Projection_Proj4_String.txt, and it is recommended that you copy the code from this file and paste it into the appropriate place in the CUSTOM COORDINATE REFERENCE SYSTEM DEFINITION window (see the flow diagram which starts on the next page), rather than entering it yourself. This will ensure that you do not encounter any problems due to the use of the incorrect Proj.4 string to define this custom projection.

Data frame with no associated projection and coordinate system

When you open a new GIS project, there will be an empty data frame. It will not have a projection or coordinate system associated with it.

1. Create a new custom projection/coordinate system

To create a new custom projection/coordinate system, click on SETTINGS on the main menu bar and select CUSTOM CRS. In the CUSTOM COORDINATE REFERENCE SYSTEM DEFINITION window which opens, click on ADD NEW CRS button (it has a green plus (+) sign on it). Next, type NORTH SEA into the NAME section about half way down the page. This means that your new custom projection/coordinate system will be called NORTH SEA. Next, enter the following text into the PARAMETERS section (**NOTE:** To ensure this is entered correctly, you are advised to copy this code from the text file called North_Sea_Projection_Proj4_String.txt, which is in the folder C:\GIS_FOR_BIOLOGISTS, and paste it into the PARAMETERS section rather than trying to enter it manually yourself):

```
+proj=tmerc +lat_0=56.5 +lon_0=-1 +k=1
     +x_0=0 +y_0=0 +ellps=WGS84
  +towgs84=0,0,0,0,0,0,0 +units=m
                +no_defs
```

This is a PROJ.4 string which tells QGIS that your custom projection/coordinate system is a transverse mercator projection (tmerc) with a latitude of origin of 56.5 degrees North and a central meridian of 1 degree west based on the WGS 1984 datum (WGS84) and that the map units are in metres (m). Once you have entered this text, check it very carefully to ensure you have got it right and then click on the ADD NEW CRS button. This will add it to the list of custom projection/coordinate systems in the top section of the window. In this section, select NEW CRS and then click the REMOVE button (it has a red minus (–) sign on it). Finally, click OK to close the CUSTOM COORDINATE REFERENCE SYSTEM DEFINITION window.

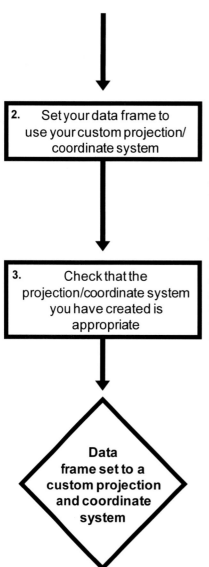

2. Set your data frame to use your custom projection/ coordinate system

In the main QGIS window, click on the CRS STATUS button in the bottom right hand corner (it is square with a dark circular design on it – see page 70 for help with where to find this button). This will open the PROJECT PROPERTIES CRS window. Click on the box next to ENABLE 'ON THE FLY' CRS TRANSFORMATION so that a cross appears in it. Next, type the name of the projection/ coordinate system into the FILTER section (in this case it will be NORTH SEA). In the COORDINATE REFERENCE SYSTEMS OF THE WORLD section, click on NORTH SEA under USER DEFINED COORDINATE SYSTEMS. This will add it to the SELECTED CRS section further down the window. Now click OK to close the PROJECT PROPERTIES CRS window.

3. Check that the projection/coordinate system you have created is appropriate

Once you have created a custom projection/ coordinate system, you need to check that it is appropriate. This involves examining how data layers look in it. For this exercise, this will be done in the next step by adding a polygon data layer of land and checking that it looks the right shape.

Data frame set to a custom projection and coordinate system

To check that you have done this step properly, click on PROJECT on the main menu bar and select PROJECT PROPERTIES. In the PROJECT PROPERTIES window which opens, click on CRS on the left hand side. Make sure that the projection/coordinate system listed in the SELECTED CRS section is called NORTH SEA and that its definition in the window below contains the following text:

+proj=tmerc +lat_0=56.5 +lon_0=-1 +k=1 +x_0=0 +y_0=0 +ellps=WGS84
+towgs84=0,0,0,0,0,0,0 +units=m +no_defs

If it does not, you will need to repeat this step to ensure that you have assigned the correct projection/coordinate system to your data frame.

STEP 2: ADD THE REQUIRED EXISTING DATA LAYERS TO YOUR GIS PROJECT:

Once the projection/coordinate system has been set for your data frame, you need to add the data layers which you will use for this exercise. These are POLYGON_GRID_NORTH_SEA_WGS84.SHP, SURVEY_TRACKS_NORTH_SEA_WGS84.SHP, BOTTLENOSE_DOLPHIN_NORTH_SEA_WGS84.SHP and LAND_NORTH_SEA_WGS84.SHP. **NOTE:** The data layer LAND_NORTH_SEA_WGS84 will not be directly used in this exercise, but is added to provide additional information for display purposes. To add these data layers to your GIS project, work through the following flow diagram:

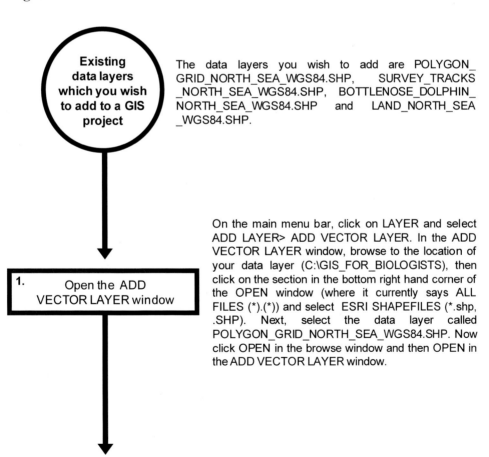

Existing data layers which you wish to add to a GIS project

The data layers you wish to add are POLYGON_GRID_NORTH_SEA_WGS84.SHP, SURVEY_TRACKS_NORTH_SEA_WGS84.SHP, BOTTLENOSE_DOLPHIN_NORTH_SEA_WGS84.SHP and LAND_NORTH_SEA_WGS84.SHP.

1. Open the ADD VECTOR LAYER window

On the main menu bar, click on LAYER and select ADD LAYER> ADD VECTOR LAYER. In the ADD VECTOR LAYER window, browse to the location of your data layer (C:\GIS_FOR_BIOLOGISTS), then click on the section in the bottom right hand corner of the OPEN window (where it currently says ALL FILES (*).(*)) and select ESRI SHAPEFILES (*.shp, .SHP). Next, select the data layer called POLYGON_GRID_NORTH_SEA_WGS84.SHP. Now click OPEN in the browse window and then OPEN in the ADD VECTOR LAYER window.

Whenever you add a data layer to a GIS project, you should always check that is has a projection/coordinate system assigned to it, and look at what this projection/coordinate system is. This is so that you know whether you will need to assign a projection/coordinate system to it, or transform it into a different projection/coordinate system before you can use it in your GIS project. To check the projection/coordinate system of your newly added data layer, right click on its name in the TABLE OF CONTENTS window and select PROPERTIES. In the LAYER PROPERTIES window which opens, click on the GENERAL tab on the left hand side and check that there is a projection/coordinate system listed in the COORDINATE REFERENCE SYSTEM section of the LAYER PROPERTIES window. For the POLYGON_GRID_NORTH_SEA_WGS84 data layer this should say:

2. Check the projection/coordinate system for the newly added data layer

USER:100000 – NORTH SEA

This tells you that this data layer is in the custom North Sea projection/coordinate system created for this study area. **NOTE:** Do not worry if the user number is different from the one provided above. What is important is that the projection/coordinate system being used for the data layer is called NORTH SEA. If the projection/coordinate system is not listed as NORTH SEA, click on the SELECT CRS button (it is at the right hand end of the COORDINATE REFERENCE SYSTEM section of the LAYER PROPERTIES window), and type NORTH SEA into the FILTER section of the COORDINATE REFERENCE SYSTEM SELECTOR window which will open. Next, select the NORTH SEA projection/coordinate system, and then click OK to close this window. Finally, click OK to close the LAYER PROPERTIES window.

3. Select how the data layer will be displayed in your GIS project

Right click on the name of the POLYGON_GRID_NORTH_SEA_WGS84 data layer in the TABLE OF CONTENTS window, and select PROPERTIES. In the LAYER PROPERTIES window, click on the STYLE tab on the left hand side. In the top left hand corner of the STYLE tab, select SINGLE SYMBOL and then click on FILL> SIMPLE FILL in the section below it. On the right hand side of the window, click on the box next to FILL STYLE and select NO BRUSH. Now, click on the box next to BORDER, select black for the colour and then click OK to close the SELECT COLOR window. Next, enter 0.25 for BORDER WIDTH and then click OK to close the LAYER PROPERTIES window. You will see that the way the POLYGON_GRID_NORTH_SEA_WGS84 data layer is displayed in the MAP window has now changed to the new settings you have just selected.

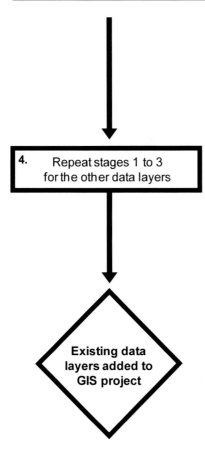

Repeat stages 1 to 3 of this step for LAND_NORTH_ SEA_WGS84.SHP, SURVEY_TRACKS_NORTH_ SEA_WGS84.SHP and BOTTLENOSE_DOLPHIN_ NORTH_SEA_WGS84. When selecting how these data layers should be displayed for LAND_ NORTH_SEA_WGS84, use dark green as the fill colour. For SURVEY_TRACKS_NORTH_SEA, use a grey line with a PEN WIDTH of 1. Finally, for BOTTLENOSE_DOLPHIN_NORTH_SEA_WGS84, select CIRCLE for the symbol, red for the colour and 3 for SIZE.

At this point, your TABLE OF CONTENTS window should look like this:

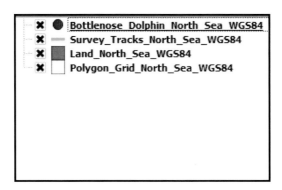

Now, right-click on the data layer called POLYGON_GRID_NORTH_ SEA_WGS84 in the TABLE OF CONTENTS window and select ZOOM TO LAYER.

The contents of your MAP window should look like this:

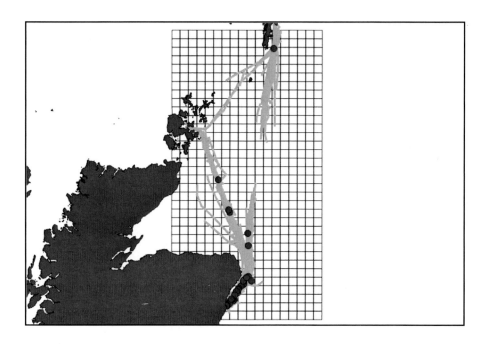

If the contents of your MAP window do not look like this, look in the TABLE OF CONTENTS window and check that you have all the required data layers added to your GIS project and that they are in the right order. If any data layers are missing, repeat step 2 until all the required data layers are present.

STEP 3: CALCULATE THE LENGTH OF SURVEY TRACKS IN EACH POLYGON GRID CELL:

The next step in this exercise is to divide the survey tracks into sections which fall in each grid cell and then calculate the lengths of all these sections. This is done using the INTERSECT tool as outlined below. Once this has been calculated, this information can be joined to the attribute table of the polygon grid data layer using a special spatial join tool in QGIS called SUM LINE LENGTHS. To carry out the required intersect and spatial join, work through the flow diagram which starts on the next page.

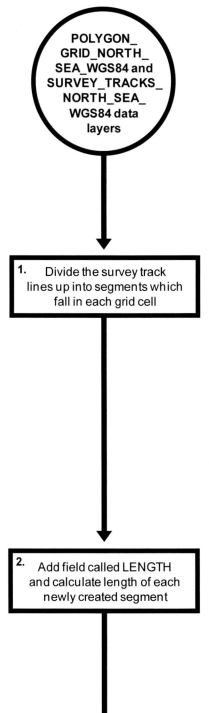

Click on VECTOR on the main menu bar and select GEOPROCESSING TOOLS> INTERSECT. In the INTERSECT tool window, select SURVEY_ TRACKS_NORTH_SEA_WGS84 from the drop down menu in the INPUT VECTOR LAYER section. Next, select the POLYGON_GRID_NORTH_SEA_ WG84 data layer from the drop down menu in the INTERSECT LAYER section. Now, click on the BROWSE button, browse to C:\GIS_FOR_ BIOLOGISTS and then enter the name SURVEY_ TRACKS_INTERSECT before clicking the SAVE button. Finally, select ADD RESULT TO CANVAS before clicking the OK button. **NOTE:** The intersect process carried out by this tool may take quite a long time to finish. Once it has finished, close the INTERSECT tool window.

Right click on the intersected line data layer's name (SURVEY_TRACKS_INTERSECT) in the TABLE OF CONTENTS window and select OPEN ATTRIBUTE TABLE. In the TABLE window, click on the TOGGLE EDITING MODE button in the top left hand corner and then click on the OPEN FIELD CALCULATOR button (it has a picture of an abacus on it and is at the right hand end of the row of buttons). In the FIELD CALCULATOR window, make sure the CREATE A NEW FIELD option is selected and then enter LENGTH in the OUTPUT FIELD NAME section and DECIMAL NUMBER (REAL) for FIELD TYPE. Next, for PRECISION enter a value of 3, and then under FUNCTIONS, double-click on GEOMETRY. Now scroll down and double-click on $LENGTH. This function will be added to the lower EXPRESSION window. Click OK to calculate the length of each segment. Next, click on the DELETE COLUMN button (it is the third button from the right at the top of the TABLE window) and select all the field names except LENGTH, before clicking the OK button. Finally, click on the SAVE EDITS button and then click on the TOGGLE EDITING MODE button to exit the editing mode before closing the TABLE window.

3. Use a SPATIAL JOIN to calculate the total length of survey effort in each grid cell of the polygon grid data layer

Click on VECTOR on the main menu bar and select ANALYSIS TOOLS> SUM LINE LENGTHS. This will open the SUM LINE LENGTHS window. In this window, select POLYGON_GRID_NORTH_ SEA_WGS84 from the drop down menu in the INPUT POLYGON VECTOR LAYER section and SURVEY_TRACKS_INTERSECT in the INPUT LINE VECTOR LAYER section. For OUTPUT SUMMED LENGTH FIELD NAME, enter LENGTH. Next, click on the BROWSE button and browse to C:\GIS_FOR_BIOLOGISTS before entering the file name POLYGON_GRID_WITH_SURVEY_EFFORT and clicking the SAVE button. Finally, select ADD RESULT TO CANVAS and click OK to run the tool. Once it has finished running, close the SUM LINE LENGTHS tool window.

4. Remove polygon grid cells with no survey effort in them

Right click on the name of your newly created polygon grid data layer (POLYGON_GRID_WITH_ SURVEY_EFFORT) and select OPEN ATTRIBUTE TABLE. In the TABLE window, click on the TOGGLE EDITING MODE button in the top left hand corner and then click on the SELECT FEATURES USING AN EXPRESION button (it has the letter E and a small yellow box on it). In the FUNCTIONS section of the SELECT BY EXPRESSION window which opens, double click on FIELDS AND VALUES and then double click on LENGTH to add it to the EXPRESSION section at the bottom of the window. Next, click on the equals symbol (=) button to add it to the EXPRESSION section, then type in the number 0. The expression should now read: "LENGTH" = 0. Click on the SELECT button and then close the SELECT BY EXPRESSION window. In the TABLE window, click on the DELETE SELECTED FEATURES button (it is beside the SAVE EDITS button). Now click the SAVE EDITS button and the TOGGLE EDITING MODE button. Finally, close the TABLE window.

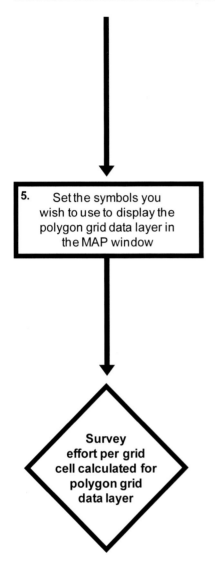

5. Set the symbols you wish to use to display the polygon grid data layer in the MAP window

Survey effort per grid cell calculated for polygon grid data layer

Right click on the name of your newly created polygon grid data layer (POLYGON_GRID_WITH_ SURVEY_EFFORT) in the TABLE OF CONTENTS window and select PROPERTIES. Next, click on the STYLE tab of the LAYER PROPERTIES window. From the drop down menu which currently says SINGLE SYMBOL, select GRADUATED. For COLUMN, select LENGTH and then for CLASSES, select 4. This will result in four ranges being added to the lowest section of the STYLE tab. In this section, double click on the section under VALUES on the first line and enter 0 for the LOWER VALUE and 10000 for the UPPER VALUE and then click OK. Double-click on the LEGEND section and enter <10,000. Click on the VALUE section for the next line down and enter 10000 for the LOWER VALUE and 50000 for the UPPER VALUE before clicking the OK button. Double-click on the LEGEND section and enter 10,000 – 50,000. Repeat this for the next line down using the range 50,000 to 100,000. Then, for the line below that, use the range 100,000 to 5,000,000. For this last range, use the legend >100,000. Finally, click OK to close the LAYER PROPERTIES window.

Once you have finished working through this flow diagram, remove the data layers SURVEY_TRACKS_INTERSECT and POLYGON_GRID_NORTH _ SEA_WGS84 from your GIS project by right-clicking on their names in the TABLE OF CONTENTS window and selecting REMOVE. Finally, change the order of the data layers by clicking on their names in the TABLE OF CONTENTS window and, while holding the mouse button down, dragging them upwards or downwards to their desired position so that your TABLE OF CONTENTS window looks like the image at the top of the next page.

233

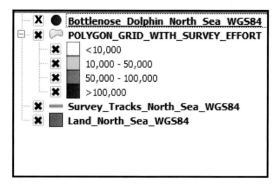

Right click on the name of the POLYGON_GRID_WITH_SURVEY_ EFFORT data layer in the TABLE OF CONTENTS window and select OPEN ATTRIBUTE TABLE. The ATTRIBUTE TABLE for the data layer POLYGON_GRID_WITH_SURVEY_EFFORT should look like this:

	CELL_ID_NO	LENGTH
0	13	1727286.155001...
1	14	0.000000000114...
2	31	2852.066079667...
3	32	1748827.893971...
4	33	0.000000000022...
5	50	96867.34583140...
6	51	1626921.326518...
7	67	1095.952889008...
8	68	32773.75566314...
9	69	614232.5843813...
10	70	1038370.511635...

Now close the TABLE window and then click on the box to the left of the data layer called SURVEY_TRACKS_NORTH_ SEA_WGS84 in the TABLE OF CONTENTS window so that the cross in it disappears and this data layer is no longer displayed.

The contents of your MAP window should now look like this:

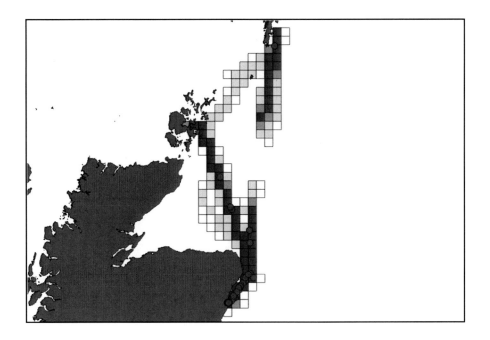

STEP 4: CALCULATE THE NUMBER OF BOTTLENOSE DOLPHINS RECORDED IN EACH GRID CELL:

The next data which must be added to the attribute table of the polygon grid data layer is information on the number of bottlenose dolphins recorded in each grid cell. This is also added using a spatial join tool (this time called JOIN ATTRIBUTES BY LOCATION in QGIS) to calculate the total number of dolphins from all sightings which fall within each grid cell. The flow diagram for this step can be found on the next page.

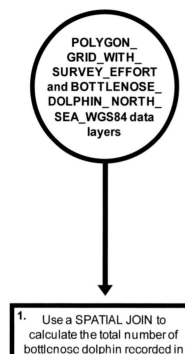

POLYGON_
GRID_WITH_
SURVEY_EFFORT
and BOTTLENOSE_
DOLPHIN_ NORTH_
SEA_WGS84 data
layers

1. Use a SPATIAL JOIN to calculate the total number of bottlenose dolphin recorded in each grid cell of the polygon grid data layer

Polygon grid data layer with a field in the attribute table with the total number of bottlenose dolphins recorded in each grid cell

Right click on the data layer BOTTLENOSE_ DOLPHIN_NORTH_SEA_WGS84 and select OPEN ATTRIBUTE TABLE. In the TABLE window, click on the TOGGLE EDITING MODE button in the top left hand corner and then click on the DELETE COLUMN button (it is the third button from the right at the top of the TABLE window). In the window that opens, select all fields except NUMBER before clicking OK. Now click the SAVE EDITS button and then the TOGGLE EDITING MODE button before closing the TABLE window.

Next, click on VECTOR on the main menu bar and select DATA MANAGEMENT TOOLS> JOIN ATTRIBUTES BY LOCATION. This will open the JOIN ATTRIBUTES BY LOCATION window. In this window, select POLYGON_GRID_WITH_SURVEY _EFFORT from the drop down menu in the TARGET VECTOR LAYER section and BOTTLENOSE_ DOLPHIN_NORTH_SEA_WGS84 in the JOIN VECTOR LAYER section. For ATTRIBUTE SUMMARY, select TAKE SUMMARY OF INTERSECTING FEATURES and then select SUM. Make sure that none of the other options are selected. Next, click on the BROWSE button and browse to C:\GIS_FOR_BIOLOGISTS before entering the file name POLYGON_GRID_WITH _SIGHTINGS and clicking the SAVE button. For OUTPUT TABLE, select KEEP ALL RECORDS (INCLUDING NON-MATCHING TARGET RECORDS). This means that all the polygon grid cells will be retained in the new data layer which will be created. Finally, click OK to run the tool. If you are asked if you would like to add the output data layer to your GIS project, click on YES. Once it has finished running, close the JOIN ATTRIBUTES BY LOCATION tool window.

At the end of this step, remove the data layer called POLYGON_ GRID_WITH_SURVEY_EFFORT by right-clicking on its name in the TABLE OF CONTENTS window and selecting REMOVE. Next, open the TABLE window for the data layer POLYGON_GRID_WITH_SIGHTINGS by right-clicking on its name in the TABLE OF CONTENTS window and selecting OPEN ATTRIBUTE TABLE. The attribute table should look like this:

	CELL_ID_NO	LENGTH	SUMNumber	COUNT
0	13	1727286.155001...	NULL	NULL
1	14	0.000000000114...	NULL	NULL
2	31	2852.066079667...	NULL	NULL
3	32	1748827.893971...	NULL	NULL
4	33	0.000000000022...	NULL	NULL
5	50	96867.34583140...	NULL	NULL
6	51	1626921.326518...	25.00000000000...	1.000000000000...
7	67	1095.952889008...	NULL	NULL
8	68	32773.75566314...	NULL	NULL
9	69	614232.5843813...	NULL	NULL
10	70	1038370.511635...	NULL	NULL
11	85	12739.52087024...	NULL	NULL
12	86	25746.89873293...	NULL	NULL
13	87	15114.75680935...	NULL	NULL
14	88	1318526.595360...	NULL	NULL
15	89	184763.6589512...	NULL	NULL
16	102	5021.708001140...	NULL	NULL
17	103	11561.47513638...	NULL	NULL
18	104	23208.71641829...	NULL	NULL
19	105	12789.16208350...	NULL	NULL
20	106	2456.970061063...	NULL	NULL

Now close the attribute table.

STEP 5: CALCULATE THE ABUNDANCE PER UNIT EFFORT IN EACH GRID CELL:

You now have all the information in the attribute table of the polygon grid data layer which you need to calculate the abundance of bottlenose dolphins per kilometre of survey effort in each grid cell. This will be calculated by adding a new field and using the FIELD CALCULATOR tool to divide the number of bottlenose dolphins recorded in each grid cell by the amount of survey effort. To do this, work through the flow diagram which starts on the next page.

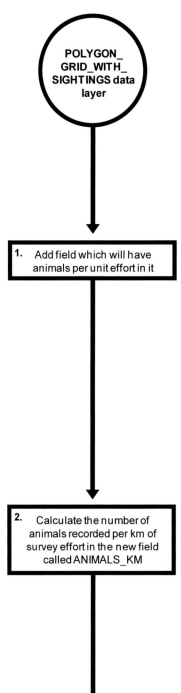

POLYGON_ GRID_WITH_ SIGHTINGS data layer

1. Add field which will have animals per unit effort in it

2. Calculate the number of animals recorded per km of survey effort in the new field called ANIMALS_KM

Right click on the name POLYGON_GRID_ WITH_SIGHTINGS in the TABLE OF CONTENTS window, and select OPEN ATTRIBUTE TABLE. Click on the TOGGLE EDITING MODE button in the top left hand corner of the TABLE window and then click on the NEW COLUMN button. In the ADD COLUMN window which will open, for NAME enter ANIMALS_KM and for TYPE select DECIMAL NUMBER (REAL). For WIDTH select 10 and for PRECISION select 6. This means that the values in the new field you are creating can have up to 10 digits in it (defined by the value used for WIDTH), with 6 of these being decimal places (defined by the value used for PRECISION). Now click OK to create the new field.

In the TABLE window, click on the OPEN FIELD CALCULATOR button (it has a picture of an abacus on it). This will open the FIELD CALCULATOR window. At the top of this window, select UPDATE EXISTING FIELD and then select ANIMALS_KM from the drop down menu immediately below it. Next, double click on FIELDS AND VALUES under FUNCTIONS, and then double click on SUMNUMBER. This will add it to the lower EXPRESSION section of the FIELD CALCULATOR window. Next, click on the divided sign (/) button and then click on the open bracket [(] button. Now double-click on LENGTH to add it to the expression. Click on the divided sign (/) button again and then type in 1000 before clicking on the close bracket [)] button. The expression should now look like this:

"SUMNumber" / ("LENGTH" / 1000)

The expression needs to be entered exactly like this (spaces and all). Now click OK. **NOTE:** The '/ 1000' term is included to convert the survey effort data from metres to kilometres. Once you have entered the expression, click on the OK button.

When you initially run the FIELD CALCULATOR tool you will find that for cells with no sightings in them, it has returned a value of NULL, yet these cells have been surveyed, so the value should be zero. To change this, click on the SELECT FEATURES USING AN EXPRESSION button (it has the letter E and a small yellow box on it). In the SELECT BY EXPRESSION window, enter the expression "ANIMALS_KM" > 0 and click on the SELECT button. This will select all the non-NULL cells. Close the SELECT BY EXPRESSION window and then click on the INVERT SELECTION button (it has a picture of two arrows on it). Once the selection has been inverted, click on the FIELD CALCULATOR button. In the FIELD CALCULATOR window, select ONLY UPDATE SELECTED FEATURES at the top, followed by UPDATE EXISTING FIELD before selecting ANIMALS_KM from the drop down menu directly below it. In the EXPRESSION section of the FIELD CALCULATOR window, type 0 and then click OK. Repeat this process with the FIELD CALCULATOR tool to remove the NULL values from the SUMNUMBER and the COUNT fields. Now click on the SAVE EDITS button and then the TOGGLE EDITING MODE button. Finally, click on the UNSELECT ALL button (it is the fifth button from the left hand side at the top of the TABLE window) and then close the TABLE window.

Right click on the name of your polygon grid data layer (POLYGON_GRID_WITH_SIGHTINGS) in the TABLE OF CONTENTS window and select PROPERTIES. Next, click on the STYLE tab of the LAYER PROPERTIES window. In the top left hand corner of the STYLE tab, click on SINGLE SYMBOL and select GRADUATED from the drop down menu that will appear. For COLUMN, select ANIMALS_KM and then for PRECISION, select a value of 4. Next, select 5 for CLASSES and then click on the CLASSIFY button. Double-click on the VALUES section of the top row of the classes created and enter 0 for the LOWER and UPPER values, then click OK. Now double-click on the LEGEND section of the same row and enter 0. Next, click on the VALUES section of the next row down and enter 0.000001 for the LOWER VALUE and 0.001 for the UPPER VALUE before clicking OK. For the LEGEND section for this row, enter <0.001. Repeat this for the next row down with a LOWER VALUE of 0.001 and and UPPER VALUE of 0.005. For the LEGEND enter 0.001 – 0.005. For the next row, use a LOWER VALUE of 0.005 and an UPPER VALUE of 0.01, and use these same values for the LEGEND. For the final range use 0.01 for the LOWER VALUE and 1 for the UPPER VALUE, and enter the LEGEND as >0.01. Now double click on the coloured box to the left of 0.0000. In the SYMBOL SELECTOR window, which will opens, click on SIMPLE FILL and select NO BRUSH for the FILL STYLE before clicking OK to close the SYMBOL SELECTOR window. Finally, click OK to close the LAYER PROPERTIES window.

3. Change the NULL values to zero values

4. Set the symbols you wish to use to display the polygon grid data layer in the MAP window

Animals per km of survey effort calculated for each grid cell in polygon grid data layer

Once you have calculated the recorded abundance of bottlenose dolphin per kilometre of survey effort per grid cell, you need to turn off all the other data layers with the exception of LAND_NORTH_SEA_WGS84 and POLYGON _GRID_WITH_SIGHTINGS so that they are no longer displayed in the MAP window. Next, open the attribute table for the data layer POLYGON_GRID_WITH_SIGHTINGS by right-clicking on its name in the TABLE OF CONTENTS window and select OPEN ATTRIBUTE TABLE. The contents of the TABLE window should look like this:

	CELL_ID_NO	LENGTH	SUMNumber	COUNT	ANIMALS_KM
0	13	1727286.155001...	0.000000000000...	0.000000000000...	0.000000
1	30	2852.066079660...	0.000000000000...	0.000000000000...	0.000000
2	31	1748827.893971...	0.000000000000...	0.000000000000...	0.000000
3	48	96867.34583135...	0.000000000000...	0.000000000000...	0.000000
4	49	1626921.326519...	25.00000000000...	1.000000000000...	0.015366
5	64	1095.952889009...	0.000000000000...	0.000000000000...	0.000000
6	65	32773.75566313...	0.000000000000...	0.000000000000...	0.000000
7	66	614232.5843811...	0.000000000000...	0.000000000000...	0.000000
8	67	1038370.511636...	0.000000000000...	0.000000000000...	0.000000
9	81	12739.52087024...	0.000000000000...	0.000000000000...	0.000000
10	82	25746.89873293...	0.000000000000...	0.000000000000...	0.000000
11	83	15114.75680935...	0.000000000000...	0.000000000000...	0.000000
12	84	1318526.595360...	0.000000000000...	0.000000000000...	0.000000
13	85	184763.6589512...	0.000000000000...	0.000000000000...	0.000000
14	97	5021.708001139...	0.000000000000...	0.000000000000...	0.000000
15	98	11561.47513638...	0.000000000000...	0.000000000000...	0.000000
16	99	23208.71641829...	0.000000000000...	0.000000000000...	0.000000
17	100	12789.16208350...	0.000000000000...	0.000000000000...	0.000000
18	101	2456.970061063...	0.000000000000...	0.000000000000...	0.000000
19	102	1333766.123049...	0.000000000000...	0.000000000000...	0.000000
20	103	58460.82254908...	0.000000000000...	0.000000000000...	0.000000

Now close the TABLE window.

Your TABLE OF CONTENTS window should look like the image at the top of the next page.

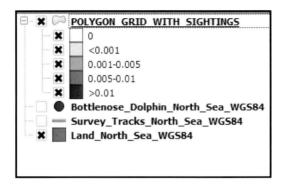

And the contents of your MAP window should look like this (**NOTE:** The exact colour of each grid cell may vary from this image due to the printing process, but the relative shading of each one should be the same):

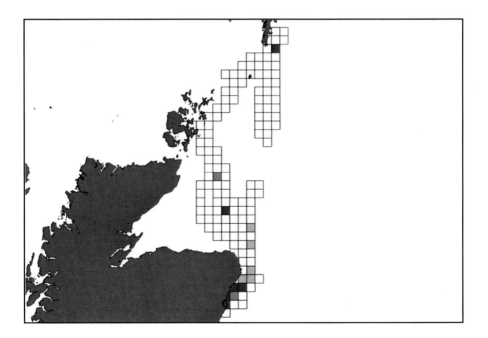

This is the end of exercise four. In this exercise, you have learned how to join information from different data layers together in two different ways. In the first, using an INTERSECT allowed you to use the information in one data layer to split the features in a second data layer (in this case, cut the lines representing survey effort at the points where they cross from one grid cell in the polygon grid data layer into another). In the second way, you used special spatial join tools to calculate both the total amount of survey effort in each cell of the polygon grid, and also the total number of bottlenose dolphin seen in each grid cell. The other ways of joining information from different data layers together, such as UNION, and EXTRACTION, work in very similar ways, but provide slightly different outputs. As a result, it is important to select the right way for joining information from different data layers together, depending on the exact outcome you wish to achieve.

Chapter Sixteen

Exercise Five: Answering Biological Research Questions With GIS Part One - Identifying Spatial Patterns In An Emerging Infectious Disease

Exercises one to four introduced you to various aspects of using GIS in biology. However, these were done from the point of view of learning how to do specific tasks, such as making a map, creating new data layers or joining data together. In contrast, when using GIS in biological research, the chances are that you will need to use all these skills, and maybe some others, to create a GIS project that will provide you with the answers that you are looking for. This means that when you come to use GIS in your research, you need to use your GIS skills in a coordinated manner. In order to give you practice in doing this, exercises five and six will take you through all the steps needed to create a GIS project to answer a specific research question of the type which you, as a biologist, are likely to need to be able to answer.

When using GIS to answer biological research questions, you need to:

- Make sure that you have clearly stated your research question.

- Identify what data layers you will need in order to answer your research question.

- Identify where you can get existing data layers from, and which data layers you will need to create for yourself.

- Ensure that you have selected an appropriate projection/coordinate system, that your data frame

243

has been set to use this projection/coordinate system *and* that all your data layers have been transformed into this projection/coordinate system before you start using them.

- Think about where you are going to store your data layers on your computer, and ensure that all your data layers are, indeed, stored there.

- Establish a protocol for backing up your GIS work, and a time table for doing this on a regular basis (ideally at the end of each session of work).

- Make sure that you take notes (lots of them!) so that you can remember what you did, what worked and what did not. Once you have finished, create a flow diagram of all the steps you took to answer your research question. This will help you remember what you did in case you ever need to do it again.

For this exercise, imagine that you have been tasked with monitoring the spread of an infectious disease across the United States of America. This disease is new to science and it causes people to become violent and attack each other. It may be a new strain of rabies, but at the moment, no one is quite sure what it is, and they do not even understand how fast it is spreading, or how the distribution of infected people is changing over time. Yet, this information is needed in order to implement a strategy to try to work out what the disease is and then how to contain it. However, what the authorities do have is information about how many people have been reported as being infected in each of the last six months in each county in the US (this information has been gathered by local hospitals and submitted to a central government agency). Your task is to use this information to create a GIS which will allow you to provide information on the distribution of incidence of the disease, and how this distribution has changed over time, which will allow those in charge to decide how best to respond. Therefore, your biological research question is: What is the spatial distribution of incidents of the new disease in each month of the last six months?

Now you have your question established, you can start thinking about how you can use GIS to answer it. There are a number of ways of assessing the distribution of things like incidence of a disease. One of the most commonly used is to simply plot all the records on a map and then draw the polygon with the smallest area which will enclose all of them. This is known as a minimum convex polygon (or MCP for short). However, this will only provide a very basic understanding of the distribution of records and it will provide no information on where the highest number of incidents have been recorded. It is also very sensitive to the presence of outliers which can greatly affect the inferred distribution and cannot easily deal with discontinuities. As a result, many people prefer to use something called a kernel density estimate (or KDE for short). A KDE builds an understanding of the spatial distribution based on the density of records and not just their presence. This allows you to identify hotspots in occurrence, and in terms of answering the current research question, this will be much more informative than a simple MCP. As a result, in this exercise, you will use KDEs to create six maps of the distribution of incidents of the new disease, one for each month you have data, and then use these to infer how the distribution has changed over this period of time. The instructions for creating the required KDEs in ArcGIS 10.3 can be found on page 246, while those for creating KDEs in QGIS 2.8.3 can be found on page 259.

If you have not already done so, before you start this exercise, you will first need to create a new folder on your C: drive called GIS_FOR_BIOLOGISTS. To do this on a computer with a Windows operating system, open Windows Explorer and navigate to your C:\ drive (this may be called Windows C:). To create a new folder on this drive, right click on the window displaying the contents of your C:\ drive and select NEW> FOLDER. This will create a new folder. Now call this folder GIS_FOR_BIOLOGISTS by typing this into the folder name to replace what it is currently called (which will most likely be NEW FOLDER). This folder, which has the address C:\GIS_FOR_BIOLOGISTS, will be used to store all files and data for the exercises in this book.

Next, you need to download the source files for the data layers outlined below from *www.gisinecology.com/GFB.htm#2*. Once you have downloaded the compressed folder containing the files, make sure that you then copy all the files it contains into the folder C:\GIS_FOR_BIOLOGISTS. **NOTE:** If you have already downloaded the compressed file from *GISinEcology.com* for any previous exercise, you will find that you have already downloaded these files, and you do not need to do this again. The data sets you will use for this exercise are:

1. **United_States_Counties.shp:** This is a point data layer where each point represents the centre of each county in continental USA (that is, all the states excluding Hawaii and Alaska). In the attribute table for this data layer, there are six fields, one for each of the last six months (month 1 is the oldest, and is the month when the first record of the new disease was noted, month 6 is the most recent). It is in the geographic projection and is based on the North American 1983 datum.

2. **United_States_Outline.shp:** This is a polygon data layer, with a single polygon that represents the continental USA. It is in the geographic projection and is based on the North American 1983 datum.

Instructions For ArcGIS 10.3 Users:

Once you have the required files downloaded into the correct folder on your computer, and you understand what is contained within each file, you can move on to creating your map. The starting point for this is a blank GIS project. To create a blank GIS project, first start the ArcGIS software by opening the ArcMap module. When it opens, you will be presented with a window which has the heading ARCMAP – GETTING STARTED. In this window you can either select an existing GIS project to work on, or create a new one. To create a new blank GIS project, click on NEW MAPS in the directory tree on the left hand side and then select BLANK MAP in the right hand section of the window. Now, click OK at the bottom of this window. This will open a new blank GIS project. (**NOTE:** If this window does not appear, you can start a new project by clicking on FILE on the main menu bar and selecting NEW. When the NEW DOCUMENT window opens, select NEW MAPS and then BLANK MAP in the order outlined above.) Once you have opened your new GIS project, the first thing you need to do is save it under a new, and meaningful, name. To do this, click on FILE on the main menu bar, and select SAVE AS. Save it as EXERCISE_FIVE in the folder C:\GIS_FOR_BIOLOGISTS.

STEP 1: SET THE PROJECTION AND COORDINATE SYSTEM OF YOUR DATA FRAME:

As with all GIS projects, the first thing which you need to do when creating a GIS to answer a specific research question is to work out what projection/ coordinate system you are going to use. You are going to use kernel density estimates to map the distribution of people infected with a disease. This means that you will have to select a projection which will allow you to accurately calculate densities. In addition, it will have to be appropriate for a study which is looking at all the continental states of the US. Of those available, one of the most appropriate is the Contiguous USA Albers projection. This projection was created for making maps of the continental USA and it uses the North American 1983 datum, which was designed to minimise spatial errors and distortions for maps of North America. As a result, this is the projection/ coordinate system which you will use for this exercise. To set your data frame to this projection/coordinate system, work through the following flow diagram:

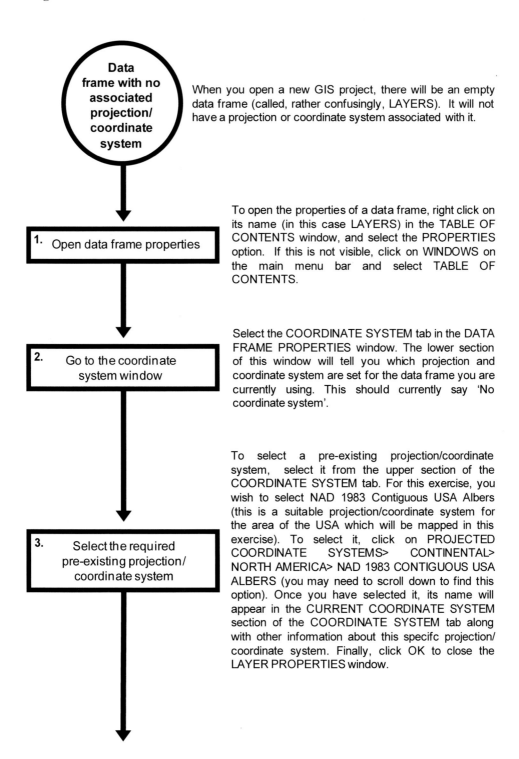

Data frame with no associated projection/ coordinate system

When you open a new GIS project, there will be an empty data frame (called, rather confusingly, LAYERS). It will not have a projection or coordinate system associated with it.

1. Open data frame properties

To open the properties of a data frame, right click on its name (in this case LAYERS) in the TABLE OF CONTENTS window, and select the PROPERTIES option. If this is not visible, click on WINDOWS on the main menu bar and select TABLE OF CONTENTS.

2. Go to the coordinate system window

Select the COORDINATE SYSTEM tab in the DATA FRAME PROPERTIES window. The lower section of this window will tell you which projection and coordinate system are set for the data frame you are currently using. This should currently say 'No coordinate system'.

3. Select the required pre-existing projection/ coordinate system

To select a pre-existing projection/coordinate system, select it from the upper section of the COORDINATE SYSTEM tab. For this exercise, you wish to select NAD 1983 Contiguous USA Albers (this is a suitable projection/coordinate system for the area of the USA which will be mapped in this exercise). To select it, click on PROJECTED COORDINATE SYSTEMS> CONTINENTAL> NORTH AMERICA> NAD 1983 CONTIGUOUS USA ALBERS (you may need to scroll down to find this option). Once you have selected it, its name will appear in the CURRENT COORDINATE SYSTEM section of the COORDINATE SYSTEM tab along with other information about this specifc projection/ coordinate system. Finally, click OK to close the LAYER PROPERTIES window.

247

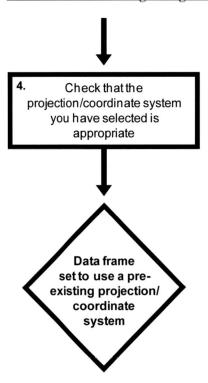

4. Check that the projection/coordinate system you have selected is appropriate

Once you have selected a projection/coordinate system, you need to check that it is appropriate. This involves examining how data layers look in it. For this exercise, this will be done in the next step by adding a series of data layers and checking that the features in them plot in the expected places and that any polygons have the expected shape.

Data frame set to use a pre-existing projection/coordinate system

To check that you have done this step properly, right click on the name of your data frame (LAYERS) in the TABLE OF CONTENTS window and select properties. Click on the COORDINATE SYSTEM tab and make sure that the contents of the CURRENT COORDINATE SYSTEM section of the window has the following text at the top of it:

NAD_1983_CONTIGUOUS_USA_ALBERS
WKID: 5070 AUTHORITY: EPSG

Projection: Albers
False_Easting: 0.0
False_Northing: 0.0
Central_Meridian: -96.0
Standard_Parallel_1: 29.5
Standard_Parallel_2: 45.5
Latitude_Of_Origin: 23
Linear Unit: Meter (1.0)

If it does not, you will need to repeat this step to ensure that you have assigned the correct projection/coordinate system to your data frame. Now click OK to close the DATA FRAME PROPERTIES window.

STEP 2: ADD THE REQUIRED EXISTING DATA LAYERS TO YOUR GIS PROJECT:

Once the projection/coordinate system has been set for your data frame, you need to add the data layers which you will use for this exercise. These are UNITED_STATES_COUNTIES.SHP and UNITED_STATES_OUTLINE.SHP. However, while both of these existing data layers use a coordinate system with the same datum as the one which you are using for this exercise (the North American 1983 datum), they are in a different projection (the geographic projection rather than the contiguous USA Albers projection). This means that they will need to be transformed into the required projection before they can be used in the GIS. This will be done as part of the process of adding existing data layers to a GIS project, and as such, a new step will be introduced into this process in comparison to earlier exercises. To do this, work through the following flow diagram:

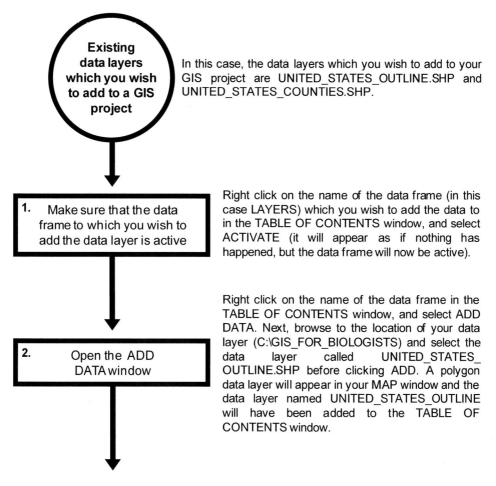

Existing data layers which you wish to add to a GIS project

In this case, the data layers which you wish to add to your GIS project are UNITED_STATES_OUTLINE.SHP and UNITED_STATES_COUNTIES.SHP.

1. Make sure that the data frame to which you wish to add the data layer is active

Right click on the name of the data frame (in this case LAYERS) which you wish to add the data to in the TABLE OF CONTENTS window, and select ACTIVATE (it will appear as if nothing has happened, but the data frame will now be active).

2. Open the ADD DATA window

Right click on the name of the data frame in the TABLE OF CONTENTS window, and select ADD DATA. Next, browse to the location of your data layer (C:\GIS_FOR_BIOLOGISTS) and select the data layer called UNITED_STATES_OUTLINE.SHP before clicking ADD. A polygon data layer will appear in your MAP window and the data layer named UNITED_STATES_OUTLINE will have been added to the TABLE OF CONTENTS window.

3. Transform the data layer into the projection/coordinate system being used for your GIS project

In the TOOLBOX window, select DATA MANAGEMENT TOOLS> PROJECTIONS AND TRANSFORMATIONS> PROJECT. This will open the PROJECT tool window. Select UNITED_ STATES_OULINE from the drop down menu in the INPUT DATASET OR FEATURE CLASS window. In the OUTPUT DATASET OR FEATURE CLASS section of the window, enter C:\GIS_FOR_BIOLOGISTS\UNITED_STATES_ OUTLINE_ALBERS.SHP.

Next, click on the button at the end of the OUTPUT COORDINATE SYSTEM section of the window to open the SPATIAL REFERENCE PROPERTIES window. To select a pre-existing projection/coordinate system, select it from the upper window. For this exercise, select PROJECTED COORDINATE SYSTEMS> CONTINENTAL> NORTH AMERICA> NAD 1983 CONTIGUOUS USA ALBERS (you may need to scroll down to find this option). Now click OK to close the SPATIAL REFERENCE PROPERTIES window, and then click OK on the PROJECT window to run the PROJECT tool. **NOTE:** If the transformed data layer is not automatically added to your GIS project, you will need to add it manually. To do this, right click on the name of the data frame (LAYERS) and select ADD data before browsing to C:\GIS_FOR_BIOLOGISTS and selecting the data layer UNITED_STATES_ OUTLINE_ALBERS.SHP.

Once you have the transformed version of your data layer in your GIS project, you can remove the original version. This is done by right-clicking on its name (UNITED_STATES_OUTLINE) in the TABLE OF CONTENTS window and selecting REMOVE.

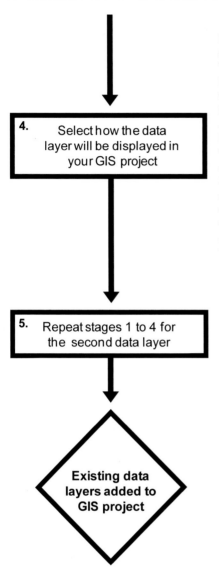

Right click on the name of the UNITED_ STATES_OUTLINE_ALBERS data layer in the TABLE OF CONTENTS window, and select PROPERTIES. In the LAYER PROPERTIES window, click on the SYMBOLOGY tab. On the left hand side, under show, select FEATURES> SINGLE SYMBOL and then click on the coloured box under SYMBOL. In the SYMBOL SELECTOR window which will open, select HOLLOW from the options on the left hand side. Click OK to close the SYMBOL SELECTOR window and then OK to close the LAYER PROPERTIES window. You will see that the way the UNITED_STATES_ OUTLINE_ALBERS data layer is displayed in the MAP window has now changed to the new settings you have just selected.

Repeat stages 1 to 4 for the data layer called UNITED_STATES_COUNTIES.SHP to add it to your GIS project and transform it into the required projection/coordinate system to create a data layer which will be called UNITED_STATES_ COUNTIES_ALBERS.SHP. For this data layer, select CIRCLE 2 for the symbol, 7 for SIZE and POINSETTIA RED for COLOR.

At the end of this step, the TABLE OF CONTENTS window should look like this:

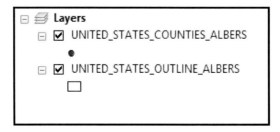

While the contents of your MAP window should look like this:

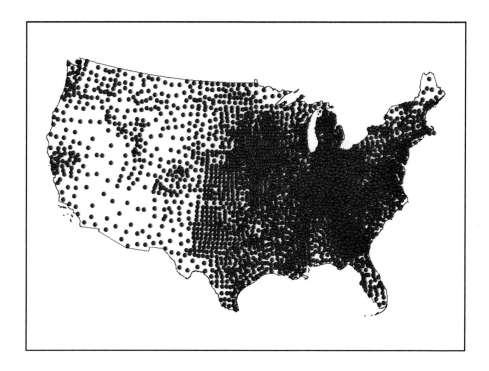

STEP 3: CREATE A KERNEL DENSITY ESTIMATE OF THE DISTRIBUTION OF THE NUMBER OF INCIDENTS OF THE DISEASE RECORDED IN EACH MONTH:

The attribute table of the point data layer of the locations of all the counties in the continental United States contains fields which have the number of recorded cases of the fictitious new disease in each of the last six months in each county. These are the values which will be used to create the kernel density estimates (KDEs) that will be used to examine how the distribution of infections has changed over this period of time. In this step, you will create six KDEs, one for each of the last six months, based on these fields from the attribute table. Within GIS, KDEs are generated using a specific kernel density estimate tool called KERNEL DENSITY. This can be done by working through the flow diagram which starts on the next page. **NOTE:** If, during this exercise you get an error message stating that you do not have an active Spatial Analyst licence, click on CUSTOMIZE on the main menu bar, select EXTENSIONS, and then make sure that the SPATIAL ANALYST extension has been activated.

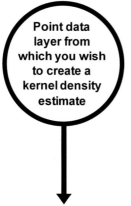

Point data layer from which you wish to create a kernel density estimate

In this exercise, the data layer which you wish to create a kernel density estimate (KDE) from is the one called UNITED_STATES_COUNTIES_ALBERS.

1. Create a kernel density estimate (KDE) raster data layer for the first month

In the TOOLBOX window, select SPATIAL ANALYST TOOLS> DENSITY> KERNEL DENSITY. In the KERNEL DENSITY window which opens, select UNITED_STATES_COUNTIES_ALBERS from the drop down menu for INPUT POINT OR POLYLINE FEATURES. For POPULATION FIELD, select MONTH_1. For OUTPUT RASTER, enter C:\GIS_FOR _BIOLOGISTS\MONTH_1 and for OUTPUT CELL SIZE (OPTIONAL) enter 100000. This will create a kernel density estimate with a cell size of 100,000m by 100,000m (or 100km by 100km). For SEARCH RADIUS (OPTIONAL), enter 100000. Now, click on ENVIRONMENTS and then, in the ENVIRONMENT SETTINGS window, click on PROCESSING EXTENT. Next, under EXTENT, select SAME AS LAYER UNITED_STATES_ OUTLINE_ALBERS. Finally, click OK to close the ENVIRONMENT SETTINGS window and then OK to run the KERNEL DENSITY tool.

When the kernel density estimate is created, it estimates values for all cells in the specified extent. This often means that they spill over into areas where you do not want estimated density values. In this example, this means estimating values which are in the sea rather than on land. Therefore, once the KDE has been created, it is often useful to remove certain cells from it. This is done by using a polygon data layer as a mask. In the case of this exercise, you will use the polygon representing the land areas of the contenential United States to mask your initial kernel density estimate.

2. Mask the kernel density estimate for the land area of the continental United States

To mask your kernel density estimate data layer, go to the TOOLBOX window and select SPATIAL ANALYST TOOLS> EXTRACTION> EXTRACT BY MASK. In the EXTRACT BY MASK tool window which will open, select MONTH_1 as the INPUT RASTER and select UNITED_STATES_ OUTLINE_ALBERS as the INPUT RASTER OR FEATURE MASK DATA. For OUTPUT RASTER, enter C:\GIS_FOR_BIOLOGISTS\MONTH_1_ MASK and then click on the OK button. Finally, right click on the raster data layer named MONTH_1 in the TABLE OF CONTENTS window and select REMOVE.

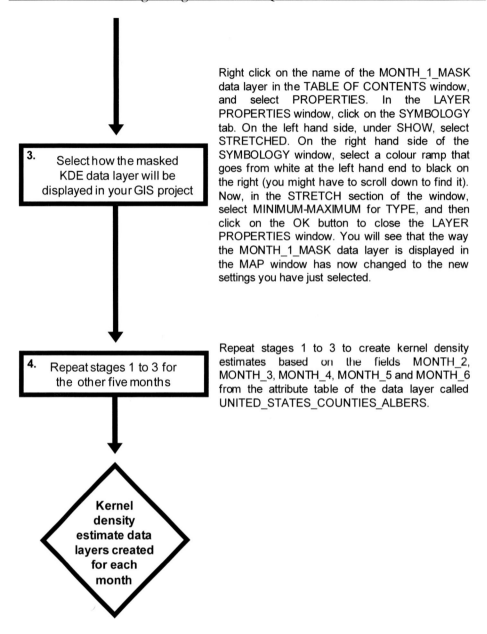

3. Select how the masked KDE data layer will be displayed in your GIS project

Right click on the name of the MONTH_1_MASK data layer in the TABLE OF CONTENTS window, and select PROPERTIES. In the LAYER PROPERTIES window, click on the SYMBOLOGY tab. On the left hand side, under SHOW, select STRETCHED. On the right hand side of the SYMBOLOGY window, select a colour ramp that goes from white at the left hand end to black on the right (you might have to scroll down to find it). Now, in the STRETCH section of the window, select MINIMUM-MAXIMUM for TYPE, and then click on the OK button to close the LAYER PROPERTIES window. You will see that the way the MONTH_1_MASK data layer is displayed in the MAP window has now changed to the new settings you have just selected.

4. Repeat stages 1 to 3 for the other five months

Repeat stages 1 to 3 to create kernel density estimates based on the fields MONTH_2, MONTH_3, MONTH_4, MONTH_5 and MONTH_6 from the attribute table of the data layer called UNITED_STATES_COUNTIES_ALBERS.

Kernel density estimate data layers created for each month

At the end of this step, click on the box next to UNITED_STATES_ COUNTIES_ALBERS in the TABLE OF CONTENTS window so that it is no longer displayed in your MAP window. Your TABLE OF CONTENTS window should now look like the image at the top of the next page.

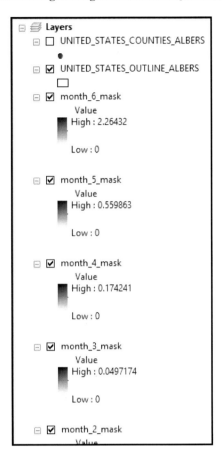

Now turn off all the raster data layers except MONTH_1_MASK so that they are not displayed in your MAP window. The contents of your MAP window should now look like this:

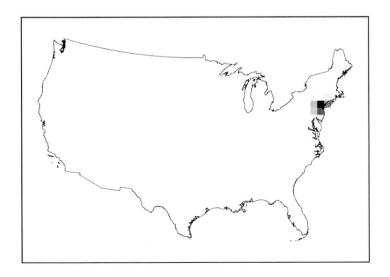

The kernel density estimate (KDE) for Month 1 shows that the disease started in New York and that at this time, all the reported cases of the disease were confined to New York City. Now, turn on the MONTH_2_MASK data layer so that it is displayed in the MAP window. It should look like this:

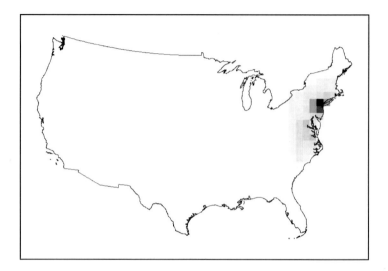

The KDE for month 2 shows that the main concentration of incidents of the new disease are still centred in New York City, but that it has also started to spread out into surrounding areas. Now, turn on the MONTH_3_MASK data layer so that it is displayed in the MAP window. It should look like this:

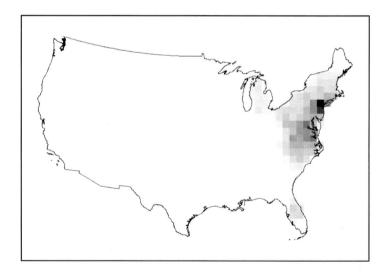

In month 3, New York still has the highest density of cases, but there is now a far greater spread into the surrounding areas. In addition, there is now a secondary location where the disease is occurring in Florida. This could represent a new, independent outbreak, but more likely it is the result of someone who was infected with the disease travelling from the area where the original outbreak occurred (around New York City) to Florida. This marks a major change in the distribution of the disease, and it would make it much harder to combat. Now, turn on the MONTH_4_MASK data layer so that it is displayed in the MAP window. It should look like this:

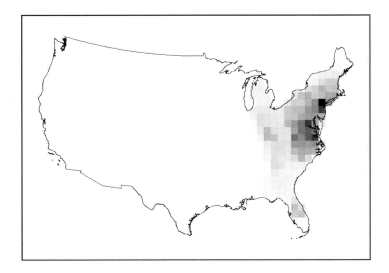

New York still has the highest density of cases in month 4, and there is still a secondary 'hotspot' of disease incidences in Florida, but the disease is now widespread throughout the eastern seaboard of the US and is spreading westward. Now, turn on the MONTH_5_MASK data layer so that it is displayed in the MAP window. It should look like this:

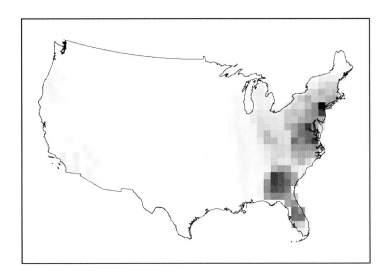

By month 5, there are at least three main concentrations of high densities of the new disease (the dark grid cells in the MONTH_5_MASK data layer), but this is set within a pattern of widespread incidents of the disease within the eastern United States. Worryingly, there are also cases now being recorded in the western United States, although the central states seem to have remained disease-free. Finally, turn on the MONTH_6_MASK data layer so that it is displayed in the MAP window. It should look like this:

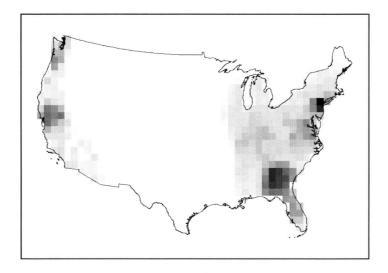

While there are still high densities of cases of the new disease in the eastern states, its spread westward in this area seems to have slowed dramatically. Instead, the main spread of the disease to new areas is happening in the western states. The central states remain disease free.

Given this pattern of spread of the new disease across the first six months of what is clearly an epidemic, three factors become clear. Firstly, the main spread of the disease seems to be at a local level, with the disease spreading out from initial incidents of infection. This can be seen first in the areas surrounding New York, and then later in Florida and on the west coast. This suggests that person-to-person transmission is the main way that this disease is spreading. Secondly, the disease is capable of making sudden leaps across long distances in short spaces of time. This is seen in month 3, where the disease suddenly turned up in Florida, and in month 5 when the disease turned up on the west coast. The most likely reason behind such sudden leaps is because someone who has contracted the disease has travelled to these new areas and taken the disease with them. This suggests that the main way to stop the disease spreading to new areas is to impose some sort of quarantine, or at least travel restrictions, to prevent this happening again. Finally, the westward spread of the disease in the eastern US seems to have come to a natural halt by month 6. This suggests that there is something going on in this area which is stopping the disease spreading further. This could be because of the geography of the local area, or the responses of local populations. Either way, identifying what has caused the spread of the disease to slow in this region may provide crucial information on how to contain and eliminate the disease in the rest of the country. Therefore, the creation of these KDEs to map the spread of the disease is providing important information which can be used to help tackle the disease and stop it spreading any further.

This is the end of exercise five, and in it you have learned how to use GIS to answer a specific biological research question. You will see from this that it is often quite straight forward to use GIS to answer such questions, and it is simply a matter of working out what you need to do to answer your specific question, and then which GIS tools you can use to get that specific answer. In this case, it was using the KERNEL DENSITY tool to generate KDEs which could be used to map the spread of a new disease, but it could just have easily been incidents of pollution, changes in human population densities, the shifting patterns of occurrence of an invasive species or, as will be seen in exercise six, the identification of biodiversity hotspots for a specific taxon.

Instructions For QGIS 2.8.3 Users:

Once you have the required files downloaded into the correct folder on your computer, and you understand what is contained within each file, you can move on to creating your GIS project for this exercise. The starting point for this is a blank GIS project. To create a blank GIS project, first open QGIS. Once it is open, click on the PROJECT menu and select SAVE AS. In the window which opens, save your GIS project as EXERCISE_FIVE in the folder C:\GIS_FOR_BIOLOGISTS.

STEP 1: SET THE PROJECTION AND COORDINATE SYSTEM OF YOUR DATA FRAME:

As with all GIS projects, the first thing which you need to do when creating a GIS to answer a specific research question is to work out what projection/ coordinate system you are going to use. You are going to use kernel density estimates to map the distribution of people infected with a disease. This means that you will have to select a projection which will allow you to accurately calculate densities. In addition, it will have to be appropriate for a study which is looking at all the continental states of the US. Of those available, one of the most appropriate is the Contiguous USA Albers projection. This projection was created for making maps of the continental USA and it uses the North American 1983 datum, which was designed to reduce artefacts for maps of North America. As a result, this is the projection/coordinate system which you will use for this exercise. To set your data frame to this projection/coordinate system, work through the following flow diagram:

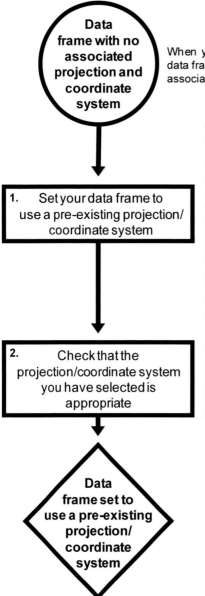

When you open a new GIS project, there will be an empty data frame. It will not have a projection or coordinate system associated with it.

In the main QGIS window, click on the CRS STATUS button in the bottom right hand corner (it is square with a dark circular design on it). This will open the PROJECT PROPERTIES CRS window. Click on the box next to ENABLE 'ON THE FLY' CRS TRANSFORMATION so that a cross appears in it. Next, type the name of the projection/ coordinate system into the FILTER section (in this case it will be USA_CONTIGUOUS_ALBERS_ EQUAL_AREA_CONIC). In the COORDINATE REFERENCE SYSTEMS OF THE WORLD section, click on USA_CONTIGUOUS_ALBERS_EQUAL_ AREA_CONIC under PROJECTED COORDINATE SYSTEMS. This will add this to the SELECTED CRS section further down the window. Now click OK to close the PROJECT PROPERTIES CRS window.

Once you have selected a pre-existing projection/ coordinate system, you need to check that it is appropriate. This involves examining how data layers look in it. For this exercise, this will be done in the next step by adding a polygon data layer of land and checking that it looks the right shape.

To check that you have done this step properly, click on PROJECT on the main menu bar and select PROJECT PROPERTIES. In the PROJECT PROPERTIES window which opens, click on CRS on the left hand side and make sure that the projection/coordinate system listed in the SELECTED CRS section is called USA_CONTIGUOUS_ALBERS_EQUAL_AREA_CONIC and that its definition in the window below contains the following text:

+proj=aea +lat_1=29.5 +lat_2=45.5 +lat_0=37.5 +lon_0=-96 +x_0=0 +y_0=0 +datum=NAD83 +units=m +no_defs

If it does not, you will need to repeat this step to ensure that you have assigned the correct projection/coordinate system to your data frame. Finally, click OK to close the PROJECT PROPERTIES window.

STEP 2: ADD THE REQUIRED EXISTING DATA LAYERS TO YOUR GIS PROJECT:

Once the projection/coordinate system has been set for your data frame, you need to add the data layers which you will use for this exercise. These are UNITED_STATES_COUNTIES.SHP   and   UNITED_STATES_OUTLINE.SHP. However, while both of these existing data layers use a coordinate system with the same datum as the one which you are using for this exercise (the North American 1983 datum), they are in a different projection (the geographic projection rather than the Contiguous USA Albers projection). This means that they will need to be transformed into the required projection before they can be used in your GIS project. This will be done as part of the process of adding existing data layers to a GIS project, and as such, a new step will be introduced into this process in comparison to earlier exercises. To do this, work through the flow diagram which starts on the next page.

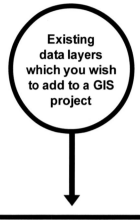

Existing
data layers
which you wish
to add to a GIS
project

In this exercise, the data layers which you wish to add to your GIS project are UNITED_STATES_ OUTLINE.SHP and UNITED_STATES_ COUNTIES.SHP.

1. Open the ADD
VECTOR LAYER window

On the main menu bar, click on LAYER and select ADD LAYER> ADD VECTOR LAYER. In the ADD VECTOR LAYER window, browse to the location of your data layer (C:\GIS_FOR_BIOLOGISTS) and click on the section in the bottom right hand corner of the OPEN window (where it currently says ALL FILES (*).(*)) and select ESRI SHAPEFILES (*.shp, .SHP). Now, select the data layer called UNITED_STATES_OUTLINE.SHP, then click OPEN in the browse window and, finally, click on OPEN in the ADD VECTOR LAYER window.

2. Transform the data layer
into the projection/coordinate
system being used for your GIS
project

In the TABLE OF CONTENTS window, right click on the name of the data layer which you have just added (UNITED_STATES_ OUTLINE) and select SAVE AS. In the SAVE VECTOR LAYER AS window, enter the file name C:/GIS_FOR_BIOLOGISTS/UNITED_ STATES_OUTLINE_ALBERS.SHP in the SAVE AS section and then select the option which starts with PROJECT CRS from the drop down menu in the CRS section. This will result in the data layer being transformed into the same projection/ coordinate system as that set for your data frame in step one. Now tick the box next to ADD SAVED FILE TO MAP and then click on the OK button.

Once you have the transformed version of your data layer in your GIS project, you can remove the original version. This is done by right-clicking on its name (UNITED_STATES_ OUTLINE) in the TABLE OF CONTENTS window and selecting REMOVE.

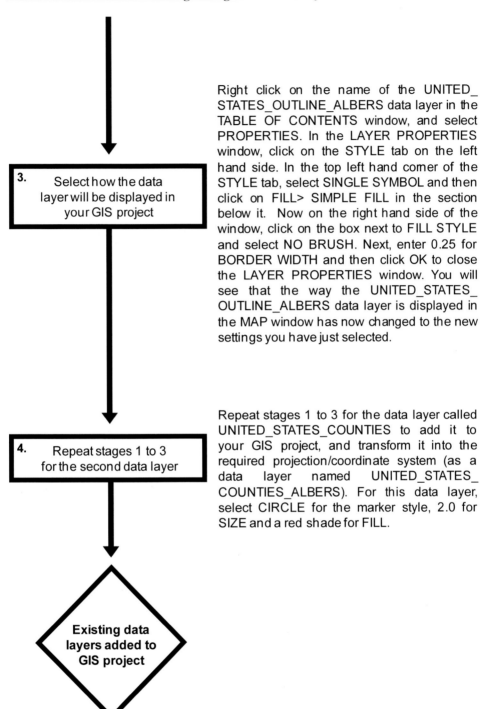

3. Select how the data layer will be displayed in your GIS project

Right click on the name of the UNITED_STATES_OUTLINE_ALBERS data layer in the TABLE OF CONTENTS window, and select PROPERTIES. In the LAYER PROPERTIES window, click on the STYLE tab on the left hand side. In the top left hand corner of the STYLE tab, select SINGLE SYMBOL and then click on FILL> SIMPLE FILL in the section below it. Now on the right hand side of the window, click on the box next to FILL STYLE and select NO BRUSH. Next, enter 0.25 for BORDER WIDTH and then click OK to close the LAYER PROPERTIES window. You will see that the way the UNITED_STATES_OUTLINE_ALBERS data layer is displayed in the MAP window has now changed to the new settings you have just selected.

4. Repeat stages 1 to 3 for the second data layer

Repeat stages 1 to 3 for the data layer called UNITED_STATES_COUNTIES to add it to your GIS project, and transform it into the required projection/coordinate system (as a data layer named UNITED_STATES_COUNTIES_ALBERS). For this data layer, select CIRCLE for the marker style, 2.0 for SIZE and a red shade for FILL.

Existing data layers added to GIS project

At the end of this step, the TABLE OF CONTENTS window should look like the image at the top of the next page.

✗	●	UNITED_STATES_COUNTIES_ALBERS
✗	☐	UNITED_STATES_OUTLINE_ALBERS

While the contents of your MAP window should look like this:

STEP 3: CREATE A KERNEL DENSITY ESTIMATE OF THE DISTRIBUTION OF THE NUMBER OF INCIDENTS OF THE DISEASE RECORDED IN EACH MONTH:

The attribute table of the point data layer of the locations of all the counties in the continental United States contains fields which have the number of recorded cases of the fictitious new disease in each of the last six months in each county. These are the values which will be used to create the kernel density estimates (KDEs) which will be used to examine how the distribution of infections has changed over this period of time. In this step, you will create six KDEs, one for each of the last six months, based on these fields from the attribute table. In QGIS, KDEs are generated using a specific kernel density estimate tool called HEATMAP. **NOTE:** To access this tool, you will need to install the HEATMAP plugin. To do this, click on PLUGINS on the main menu bar and select MANAGE AND INSTALL plugins. In the window that opens, click on the ALL tab on the left hand side and then locate the HEATMAP plugin from the list provided before clicking INSTALL. If it is already installed, make sure that it has been activated (i.e. that a cross or tick is present in the small box next to its name). Now close the PLUGINS window and you will now be able to access the HEATMAP plugin through the RASTER menu as described in the flow diagram on the next page.

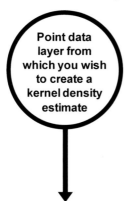

Point data layer from which you wish to create a kernel density estimate

In this case, the data layer which you wish to create a kernel density estimate (KDE) from is the one called UNIITED_ STATES_COUNTIES_ALBERS.

On the main menu bar, select RASTER> HEATMAP> HEATMAP. This will open the HEATMAP PLUGIN window. In this window, select UNITED_STATES_ COUNTIES_ALBERS as the INPUT POINT LAYER. Next, for OUTPUT RASTER, enter C:/GIS_FOR_ BIOLOGISTS/MONTH_1.TIF, and select GEOTIFF for OUTPUT FORMAT. For RADIUS, enter 100000 and then click on the box next to ADVANCED. This will activate the ADVANCED settings. In the ADVANCED settings, enter 100000 for CELL SIZE X and for CELL SIZE Y. This means that the resulting raster data layer will have cells which are 100,000m by 100,000m, or 100km by 100km. Next, select QUARTIC (BIWEIGHT) for KERNEL SHAPE and then tick the box next to USE WEIGHT FROM FIELD. Once it is active, select MONTH_1 from the drop down menu beside this option. Finally, click OK to run the HEATMAP tool and create your kernel density estimate for month 1.

1. Create a kernel density estimate (KDE) raster data layer for the first month

When the kernel density estimate is created, it estimates values for all cells in the specified extent. This often means that they spill over into areas where you do not want estimated density values. In this example, this means estimating values which are in the sea rather than on land. Therefore, once the KDE has been created, it is often useful to remove certain cells from it. This is done by using a polygon data layer as a mask. In the case of this exercise, you will use the polygon representing the land areas of the continential United States to mask your initial kernel density estimate.

2. Mask the kernel density estimate for the land area of the continental United States

To mask your kernel density estimate data layer, click on RASTER on the main menu bar and select EXTRACTION> CLIPPER. In the CLIPPER window, select MONTH_1 as the INPUT FILE (RASTER) and then enter C:/GIS_FOR_BIOLOGISTS/MONTH_1_ MASK.TIF in the OUTPUT FILE section. Next, tick the box next to NO DATA VALUE and select 0 from the drop down menu beside it. Now, in the CLIPPING MODE section, click on the MASK LAYER option and select UNITED_STATES_OUTLINE_ALBERS as the MASK LAYER. Next, make sure that the box beside LOAD INTO CANVAS WHEN FINISHED is ticked and then click on the OK button. Once the tool has finished running, click on the CLOSE button to close the CLIPPER window. Finally, right click on the raster data layer named MONTH_1 in the TABLE OF CONTENTS window and select REMOVE.

265

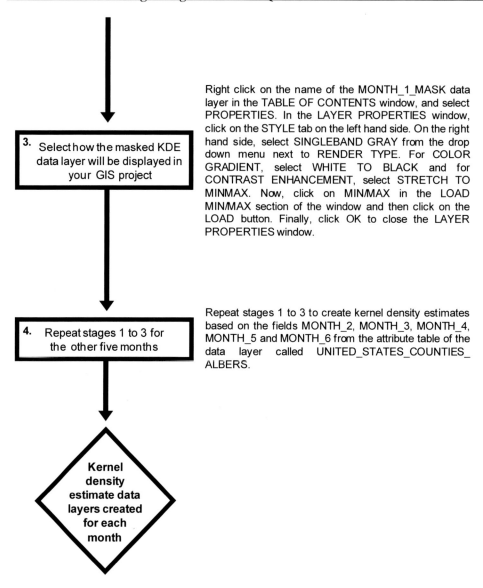

3. Select how the masked KDE data layer will be displayed in your GIS project

Right click on the name of the MONTH_1_MASK data layer in the TABLE OF CONTENTS window, and select PROPERTIES. In the LAYER PROPERTIES window, click on the STYLE tab on the left hand side. On the right hand side, select SINGLEBAND GRAY from the drop down menu next to RENDER TYPE. For COLOR GRADIENT, select WHITE TO BLACK and for CONTRAST ENHANCEMENT, select STRETCH TO MINMAX. Now, click on MIN/MAX in the LOAD MIN/MAX section of the window and then click on the LOAD button. Finally, click OK to close the LAYER PROPERTIES window.

4. Repeat stages 1 to 3 for the other five months

Repeat stages 1 to 3 to create kernel density estimates based on the fields MONTH_2, MONTH_3, MONTH_4, MONTH_5 and MONTH_6 from the attribute table of the data layer called UNITED_STATES_COUNTIES_ALBERS.

Kernel density estimate data layers created for each month

At the end of this step, click on the box next to UNITED_STATES_COUNTIES_ALBERS in the TABLE OF CONTENTS window so that it is no longer displayed in your MAP window, and then click on the name of the data layer UNITED_STATES_OUTLINE_ALBERS and drag it to the top of the TABLE OF CONTENTS window so that it is displayed on top of the kernel density estimate data layers. Your TABLE OF CONTENTS window should now look like the image at the top of the next page.

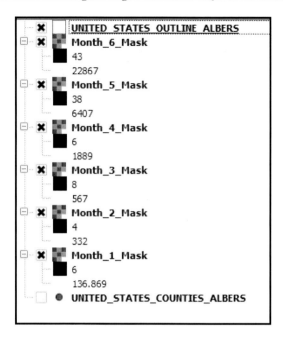

Now turn off all the raster data layers except MONTH_1_MASK so that they are not displayed in your MAP window. The contents of your MAP window should now look like this:

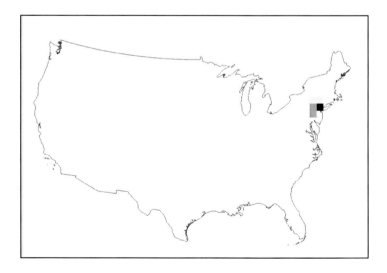

The kernel density estimate (KDE) for Month 1 shows that the disease started in New York and that at this time, all the reported cases of the disease were confined to New York City. Now, turn on the MONTH_2_MASK data layer so that it is displayed in the MAP window. It should look like this:

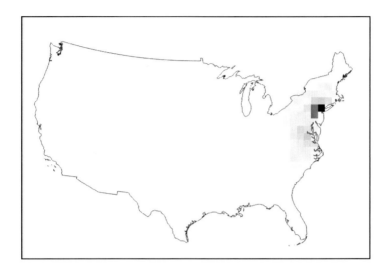

The KDE for month 2, shows that the main concentration of incidents of the new disease are still centred in New York City, but that it has also started to spread out into surrounding areas. Now, turn on the MONTH_3_MASK data layer so that it is displayed in the MAP window. It should look like this:

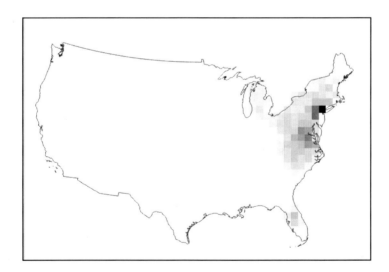

In month 3, New York still has the highest density of cases, but there is now a far greater spread into the surrounding areas. In addition, there is a secondary location where the disease is now occurring in Florida. This could represent a new, independent outbreak, but more likely it is the result of someone who was infected with the disease travelling from the area where the original outbreak occurred (around New York City) to Florida. This marks a major change in the distribution of the disease, and it would make it much harder to combat. Now, turn on the MONTH_4_MASK data layer so that it is displayed in the MAP window. It should look like this:

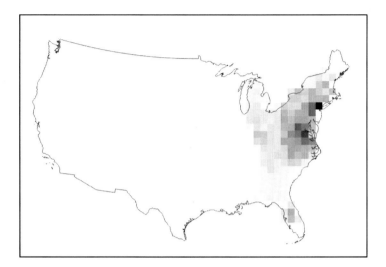

New York still has the highest density of cases in month 4, and there is still a secondary 'hotspot' of disease incidences in Florida, but the disease is now widespread throughout the eastern seaboard of the US and is spreading westward. Now, turn on the MONTH_5_MASK data layer so that it is displayed in the MAP window. It should look like this:

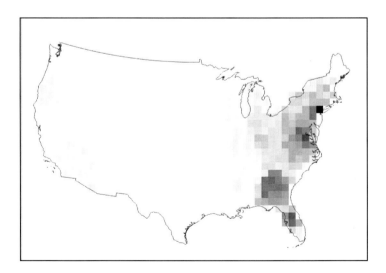

By month 5, there are at least three main concentrations of high densities of the new disease (the dark grid cells in the MONTH_5_MASK data layer), but this is set within a pattern of widespread incidents of the disease within the eastern United States. Worryingly, there are also cases now being recorded in the western United States, although the central states seem to have remained disease-free. Finally, turn on the MONTH_6_MASK data layer so that it is displayed in the MAP window. It should look like this:

While there are still high densities of cases of the new disease in the eastern states, its spread westward in this area seems to have slowed dramatically. Instead, the main spread of the disease to new areas is happening in the western states. The central states remain disease free.

Given this pattern of spread of the new disease across the first six months of what is clearly an epidemic, three factors become clear. Firstly, the main spread of the disease seems to be at a local level, with the disease spreading out from initial incidents of infection. This can be seen first in the areas surrounding New York, and then later in Florida and on the west coast. This suggests that person-to-person transmission is the main way that this disease is spreading. Secondly, the disease is capable of making sudden leaps across long distances in short spaces of time. This is seen in month 3, where the disease suddenly turned up in Florida, and in month 5 when the disease turned up on the west coast. The most likely reason behind such sudden leaps is because someone who has contracted the disease has travelled to these new areas and taken the disease with them. This suggests that the main way to stop the disease spreading to new areas is to impose some sort of quarantine, or at least travel restrictions, to prevent this happening again. Finally, the westward spread of the disease in the eastern US seems to have come to a natural halt by month 6. This suggests that there is something going on in this area which is stopping the disease spreading further. This could be because of the geography of the local area, or the responses of local populations. Either way, identifying what has caused the spread of the disease to slow in this region may provide crucial information on how to contain and eliminate the disease in the rest of the country. Therefore, the creation of these KDEs to map the spread of the disease is providing important information which can be used to help tackle the disease and stop it spreading any further.

This is the end of exercise five, and in it you have learned how to use GIS to answer a specific biological research question. You will see from this that it is often quite straight forward to use GIS to answer such questions, and it is simply a matter of working out what you need to do to answer your specific question, and then which GIS tools you can use to get that specific answer. In this case, it was using the HEATMAP tool to generate KDEs which could be used to map the spread of a new disease, but it could just have easily been incidents of pollution, changes in human population densities, the shifting patterns of occurrence of an invasive species or, as will be seen in exercise six, the identification of biodiversity hotspots for a specific taxon.

--- *Chapter Seventeen* ---

Exercise Six: Answering Biological Research Questions With GIS Part Two: Identifying Patterns In Species Diversity

In exercise five, you worked through an example of how GIS can be used to answer a biological research question about how the distribution of people infected with a fictitious disease changed over time. In this exercise, you will investigate a second biological research question regarding identifying hotspots of biodiversity for a group of poorly known whales in the North Atlantic Ocean. The beaked whales are a family of twenty-two known species of whales that spend most of their lives in the deep ocean where they often forage at depths of 500m (~1,500ft) or more. Their feeding dives can last for more than an hour, and they spend up to 67% of their time foraging at depth. Little is known about their biology, but we do know that they are impacted by a variety of human activities. This includes the ingestion of plastic bags, being hunted for meat, being accidently caught in fishing nets and climate change. However, the one activity which affects beaked whales that has received the most publicity in recent years is noise pollution. Specifically, there have been a number of mass strandings which have occurred at the same time as, or shortly after, naval exercises which have used mid-frequency sonars.

One approach to mitigating the effects of such activities on beaked whales is to identify 'hotspots' in their distribution and then carefully manage human activities in these specific areas. There are a number of different ways that hotspots can be identified, but one of the most common is to look for areas which have high levels of species diversity. This, then, is your research question: where is there a high diversity of beaked whale species in the North Atlantic? To answer this research question, you will create a raster data layer which contains information about the diversity of beaked whale species in a grid of 50km by 50 km grid cells. This will be done using a tool called a raster calculator. Raster calculator tools are very useful because they allow you to create new raster data layers by applying mathematical functions to the values in over-lapping cells in other existing data layers.

There are six species of beaked whale which are known to regularly occur in the North Atlantic Ocean. These are Sowerby's beaked whale, True's beaked whale, Gervais' beaked whale, Blainville's beaked whale, Cuvier's beaked whale and the northern bottlenose whale. For this exercise, you will be provided with feature data layers which contain polygons representing the current range of these six species in the North Atlantic. Once you have created your GIS project for this exercise, you will convert these polygons into raster data layers which have a value of 1 for grid cells which fall within the range of a species, and a value of 0 for grid cells which do not. These raster data layers will then be added together using a raster calculator tool to create a new raster data layer where each cell has a value that represents the number of beaked whale species recorded in it.

The main issue which you are likely to encounter when conducting this type of analysis is related to your choice of projection/coordinate system. There are two related problems here. Firstly, you are going to be working at the scale of a whole ocean. This means that you need to select a projection/coordinate system which will not distort any features you are interested in across this whole area. You faced a similar problem in exercise five when investigating the spread of a disease across the United States of America, but in that instance you were able to select a pre-existing projection/coordinate system that has already been established to allow you to do this for the continental United States. This, then, is the second issue, there is no such standard projection/coordinate system for the North Atlantic Ocean. This means that you would need to select and then create your own custom projection/coordinate system that is suitable for your region of interest. In this case, you will use an Albers equal-area conic projection. This is the same type as the one you used in exercise five, but rather than being centred on the continental United States, you will use one which is centred on the North Atlantic. This is done by using -30 degrees as the central meridian, 55 degrees north as the reference parallel, and 45 and 65 degrees north as its standard parallels. In addition, you will use the WGS 1984 datum (which is widely used in the marine environment) rather than the North American 1983 datum which was used in exercise five. This projection/coordinate system will be created in step one, and the polygon data layers will then be transformed into this projection/coordinate system in step two.

The instructions for creating a species diversity raster data layer from information on species ranges in ArcGIS 10.3 can be found on page 275, while those for QGIS 2.8.3 can be found on page 292.

If you have not already done so, before you start this exercise, you will first need to create a new folder on your C: drive called GIS_FOR_BIOLOGISTS. To do this on a computer with a Windows operating system, open Windows Explorer and navigate to your C:\ drive (this may be called Windows C:). To create a new folder on this drive, right click on the window displaying the contents of your C:\ drive and select NEW> FOLDER. This will create a new folder. Now call this folder GIS_FOR_BIOLOGISTS by typing this into the folder name to replace what it is currently called (which will most likely be NEW FOLDER). This folder, which has the address C:\GIS_FOR_BIOLOGISTS, will be used to store all files and data for the exercises in this book. Next, you need to download the source files for the data layers outlined below from *www.gisinecology.com/GFB.htm#2*. Once you have downloaded the compressed folder containing the files, make sure that you then copy all the files it contains into the folder C:\GIS_FOR_BIOLOGISTS. **NOTE:** If you have already downloaded the compressed file from *GISinEcology.com* for any previous exercise, you will find that you have already downloaded these files, and you do not need to do this again. The data sets you will use for this exercise are:

1. **Countries_Of_The_World.shp:** This is a polygon data layer of all the countries in the world. It was originally downloaded from *www.thematicmapping.org.* This will be used simply to provide information about where areas of high beaked whale diversity fall in relation to the nearest areas of land. It is in the geographic projection and is based on the WGS 1984 datum.

2. **North_Atlantic_Ocean.shp:** This is a polygon data layer of the North Atlantic Ocean. It will be used to ensure that the final raster data layer only contains information for this ocean and not any other parts of the world. It is in the geographic projection and is based on the WGS 1984 datum.

3. **Sowerbys_Beaked_Whale_Range:** This is a polygon data layer, with a single polygon that represents the global range of Sowerby's beaked whale. It is in the geographic projection and is based on the WGS 1984 datum.

4. **Trues_Beaked_Whale_Range:** This is a polygon data layer, with a single polygon that represents the global range of True's beaked whale. It is in the geographic projection and is based on the WGS 1984 datum.

5. **Gervais_Beaked_Whale_Range:** This is a polygon data layer, with a single polygon that represents the global range of Gervais beaked whale. It is in the geographic projection and is based on the WGS 1984 datum.

6. **Blainvilles_Beaked_Whale_Range:** This is a polygon data layer, with a single polygon that represents the global range of Blainville's beaked whale. It is in the geographic projection and is based on the WGS 1984 datum.

7. **Cuviers_Beaked_Whale_Range:** This is a polygon data layer, with a single polygon that represents the global range of Cuvier's beaked whale. It is in the geographic projection and is based on the WGS 1984 datum.

8. **Northern_Bottlenose_Whale_Range:** This is a polygon data layer, with a single polygon that represents the global range of the northern bottlenose whale. It is in the geographic projection and is based on the WGS 1984 datum.

The range maps for these beaked whale species were created as part of the *Digital Beaked Whale Atlas Project*. More information about how they were created and exactly what they represent can be found at *tinyurl.com/GFB-Link8*. This digital atlas also contains range maps for all other known beaked whale species.

Instructions For ArcGIS 10.3 Users:

Once you have all the required files downloaded into the correct folder on your computer, and you understand what is contained within each file, you can move on to creating your GIS project. The starting point for this is a blank GIS project. To create a blank GIS project, first start the ArcGIS software by opening the ArcMap module. When it opens, you will be presented with a window which has the heading ARCMAP – GETTING STARTED. In this window you can either select an existing GIS project to work on, or create a new one. To create a new blank GIS project, click on NEW MAPS in the directory tree on the left hand side and then select BLANK MAP in the right hand section of the window. Now, click OK at the bottom of this window. This will open a new blank GIS project. (**NOTE:** If this window does not appear, you can start a new project by clicking on FILE on the main menu bar, and selecting NEW. When the NEW DOCUMENT window opens, select NEW MAPS and then BLANK MAP in the order outlined above.) Once you have opened your new GIS project, the first thing you need to do is save it under a new, and meaningful, name. To do this, click on FILE on the main menu bar, and select SAVE AS. Save it as EXERCISE_SIX in the folder C:\GIS_FOR_BIOLOGISTS.

STEP 1: SET THE PROJECTION AND COORDINATE SYSTEM OF YOUR DATA FRAME:

As with all GIS projects, the first thing which you need to do when creating a GIS to answer a specific research question is to work out what projection/coordinate system you are going to use. You are going to create a raster data layer of beaked whale species diversity which will cover the whole North Atlantic Ocean. This means that you need to use a projection/coordinate system which will not cause any major biases over such a large area. One suitable projection is the Albers equal-area conic projection, and you will create a custom version of this projection centred on the middle of the North Atlantic. In addition, you will use the WGS 1984 datum for the coordinate system. To set your data frame to this custom projection/coordinate system, work through the following flow diagram:

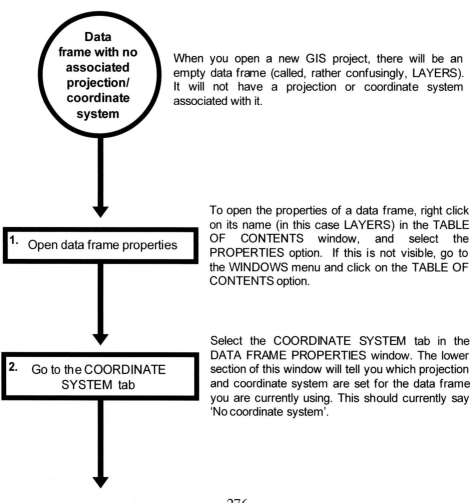

Data frame with no associated projection/coordinate system

When you open a new GIS project, there will be an empty data frame (called, rather confusingly, LAYERS). It will not have a projection or coordinate system associated with it.

1. Open data frame properties

To open the properties of a data frame, right click on its name (in this case LAYERS) in the TABLE OF CONTENTS window, and select the PROPERTIES option. If this is not visible, go to the WINDOWS menu and click on the TABLE OF CONTENTS option.

2. Go to the COORDINATE SYSTEM tab

Select the COORDINATE SYSTEM tab in the DATA FRAME PROPERTIES window. The lower section of this window will tell you which projection and coordinate system are set for the data frame you are currently using. This should currently say 'No coordinate system'.

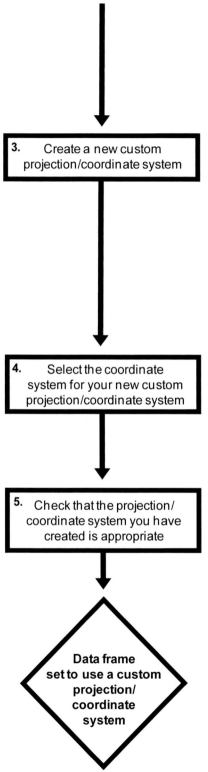

3. Create a new custom projection/coordinate system

4. Select the coordinate system for your new custom projection/coordinate system

5. Check that the projection/coordinate system you have created is appropriate

Data frame set to use a custom projection/coordinate system

To create a new custom projection/coordinate system, press the ADD COORDINATE SYSTEM button that can be found towards the top right hand corner of the COORDINATE SYSTEM tab (it has a picture of a globe on it) and select NEW> PROJECTED COORDINATE SYSTEM. This will open the NEW PROJECTED COORDINATE SYSTEM window. In the upper NAME window, type in NORTH ATLANTIC. In the PROJECTION portion of the window, select the name of the appropriate type of coordinate system from the drop down menu (in the lower NAME section). For this exercise, select ALBERS. Next, type in the values you wish to use for the parameters. For CENTRAL_MERIDIAN, enter -30. For STANDARD_PARALLEL_1, enter 25 and for STANDARD_PARALLEL_2, enter 65. For LATITUDE_OF_ORIGIN, enter 45. Leave all other sections of the window with their default settings.

In the GEOGRAPHIC COORDINATE SYSTEM section of the PROJECTED COORDINATE SYSTEM window it should say, by default, NAME: GCS_WGS_1984. If it doesn't, click on the CHANGE button and type WGS 1984 into the SEARCH box in the window that appears and press the ENTER key on your keyboard. Select WORLD> WGS 1984, and click the OK button. Now click the OK button in the NEW PROJECTED COORDINATE SYSTEM window. Finally, click OK in the DATA FRAME PROPERTIES window.

Once you have created a custom projection/ coordinate system, you need to check that it is appropriate. This involves examining how data layers look in it. For this exercise, this will be done in the next step by adding a polygon data layer of land and checking that it looks the right shape.

To check that you have done this step properly, right click on the name of your data frame (LAYERS) in the TABLE OF CONTENTS window and select PROPERTIES. Click on the COORDINATE SYSTEM tab and make sure that the contents of the CURRENT COORDINATE SYSTEM section of the window has the following text at the top of it:

North Atlantic
Authority: Custom

Projection: Albers
False_Easting: 0.0
False_Northing: 0.0
Central_Meridian: -30.0
Standard_Parallel_1: 25.0
Standard_Parallel_2: 65.0
Latitude_Of_Origin: 45.0
Linear Unit: Meter (1.0)

If it does not, you will need to repeat this step to ensure that you have assigned the correct projection/coordinate system to your data frame.

Now click OK to close the DATA FRAME PROPERTIES window.

STEP 2: ADD THE REQUIRED DATA LAYERS TO YOUR GIS PROJECT:

Once the projection/coordinate system has been set for your data frame, you need to add the data layers which you will use for this exercise. These data layers are COUNTRIES_OF_THE_WORLD.SHP, NORTH_ATLANTIC_OCEAN.SHP and the six polygon data layers which contain the information about the ranges of the six species of beaked whale regularly found in the North Atlantic. However, while all of these existing data layers use a coordinate system with the same datum as the one which you are using for this exercise (WGS 1984), they are in a different projection (the geographic projection rather than the custom Albers projection which will be used for this exercise). This means that they will need to be transformed into the required projection before they can be used in your GIS project. As was the case in exercise five, this will be done as part of the process of adding existing data layers to a GIS project. To add the required existing data layers to your GIS project, work through the following flow diagram:

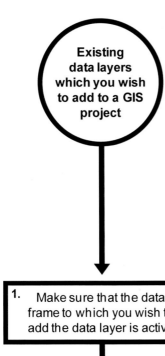

The data layers you wish to add are COUNTRIES_OF_THE_WORLD.SHP, NORTH_ATLANTIC_OCEAN.SHP, SOWERBYS_BEAKED_WHALE_RANGE.SHP, BLAINVILLES_BEAKED_WHALE_RANGE.SHP, GERVAIS_BEAKED_WHALE_RANGE.SHP, TRUES_BEAKED_WHALE_RANGE.SHP, CUVIERS_BEAKED_WHALE.SHP and NORTHERN_BOTTLENOSE_WHALE.SHP.

1. Make sure that the data frame to which you wish to add the data layer is active

Right click on the name of the data frame (in this case LAYERS) which you wish to add the data to in the TABLE OF CONTENTS window, and select ACTIVATE (it will appear as if nothing has happened, but the data frame will now be active).

2. Open the ADD DATA window

Right click on the name of the data frame (LAYERS) in the TABLE OF CONTENTS window, and select ADD DATA. Browse to the location of your data layer (C:\GIS_FOR_BIOLOGISTS) and select the data layer called COUNTRIES_OF_THE_WORLD.SHP before clicking ADD. A polygon data layer will appear in your MAP window and the data layer named COUNTRIES_OF_THE_WORLD will have been added to the TABLE OF CONTENTS window.

3. Transform the data layer into the projection/coordinate system being used for your GIS project

Right click on the name of the data layer you have just added in the TABLE OF CONTENTS window (COUNTRIES_OF_THE_WORLD) and select DATA> EXPORT DATA. In The EXPORT DATA window, select ALL FEATURES in the EXPORT section and then under USE THE SAME COORDINATE SYSTEM AS, select THE DATA FRAME by clicking the circle beside it. By selecting this option, it allows you to create a new data layer which is in the same projection/ coordinate system that the data frame was set to use in step one. **NOTE:** This is an alternative way to transform the projection/coordinate system of a data layer than using the PROJECT tool as was done in exercise five. Finally, in the OUTPUT DATASET OR FEATURE CLASS section of the window, enter C:\GIS_FOR_BIOLOGISTS\ COUNTRIES_OF_THE_WORLD_ALBERS, and then click OK to export the data layer. If asked, click YES to add the new data layer you have just created to your GIS project.

Once you have the transformed version of your data layer in your GIS project, you can remove the original version. This is done by right-clicking on its name (COUNTRIES_OF_THE_WORLD) in the TABLE OF CONTENTS window and selecting REMOVE.

4. Select how the data layer will be displayed in your GIS project

Right click on the name of the COUNTRIES_ OF_THE_WORLD_ALBERS data layer in the TABLE OF CONTENTS window, and select PROPERTIES. In the LAYER PROPERTIES window, click on the SYMBOLOGY tab. On the left hand side, under SHOW, select FEATURES> SINGLE SYMBOL and then click on the coloured box under SYMBOL. In the SYMBOL SELECTOR window which will open, select SPRUCE GREEN for FILL COLOR. Now click OK to close the SYMBOL SELECTOR window and then OK to close the LAYER PROPERTIES window. You will see that the way the COUNTRIES_OF_ THE_WORLD_ALBERS data layer is displayed in the MAP window has now changed to the new settings you have just selected.

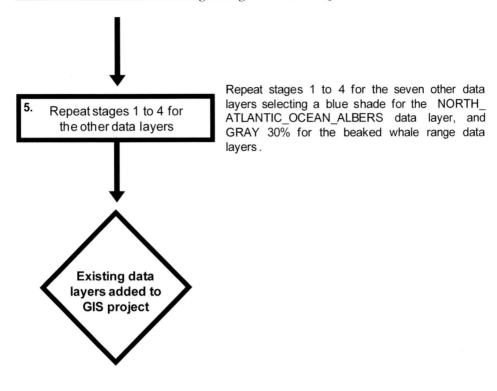

5. Repeat stages 1 to 4 for the other data layers

Repeat stages 1 to 4 for the seven other data layers selecting a blue shade for the NORTH_ ATLANTIC_OCEAN_ALBERS data layer, and GRAY 30% for the beaked whale range data layers.

Existing data layers added to GIS project

At the end of this step, the TABLE OF CONTENTS window should look like this:

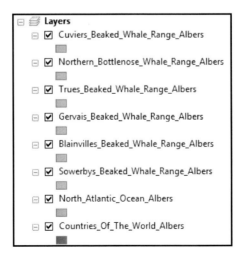

Now, in the TABLE OF CONTENTS window, right click on the data layer called NORTH_ATLANTIC_OCEAN_ALBERS and select ZOOM TO LAYER. The contents of your MAP window should now look like this:

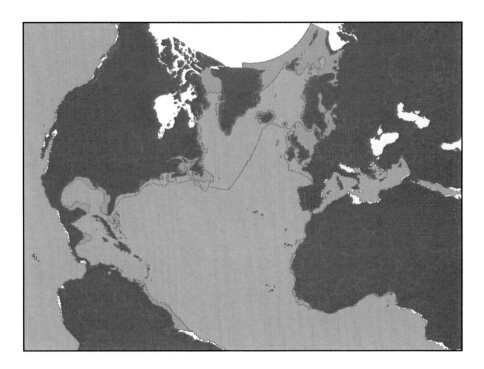

STEP 3: CREATE RASTER DATA LAYERS OF THE RANGES OF THE SIX NORTH ATLANTIC BEAKED WHALE SPECIES:

Once you have the required data layers added to your GIS project and then transformed into the same projection/coordinate system as your data frame, you next need to convert each of the species range polygon data layers into a raster data layer. This is so that you can use the RASTER CALCULATOR tool to add up the number of beaked whale species which occur in each grid cell. Creating a raster data layer of species range from a polygon is relatively straight forward and can be done with the POLYGON TO RASTER conversion tool. However, once this data layer has been created, you need to do some further processing. This processing involves making sure that all the grid cells which are outside of the range of a specific species have a value of zero (which is required for the raster calculator tool to work properly), and then masking the raster data layer to remove any grid cells which do not fall in the North Atlantic Ocean. To make the species raster data layers, work through the flow diagram which starts on the next page.

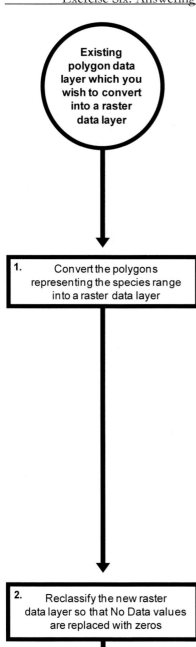

Existing polygon data layer which you wish to convert into a raster data layer

1. Convert the polygons representing the species range into a raster data layer

2. Reclassify the new raster data layer so that No Data values are replaced with zeros

In the TOOLBOX window, select CONVERSION TOOLS> TO RASTER> POLYGON TO RASTER. In the POLYGON TO RASTER window, select the polygon data layer called SOWERBYS_BEAKED_ WHALE_RANGE_ALBERS in the INPUT FEATURES section using the drop down menu. Select the field called EXPT_RANGE using the drop down menu in the VALUE FIELD section. Next, type C:\GIS_FOR_ BIOLOGISTS\SBW_RANGE in the OUTPUT RASTER DATASET window. In the CELL ASSIGNMENT TYPE (OPTIONAL) section, select CELL_CENTER, and then type the number 50000 into the CELLSIZE (OPTIONAL) section. Next, click on the ENVIRONMENTS button at the bottom of the window. In the ENVIRONMENT SETTINGS window which will open, click on PROCESSING EXTENT. In the EXTENT section of the window that will appear, select SAME AS LAYER NORTH_ATLANTIC_OCEAN _ALBERS. This will match the extent of the new raster data layer to the extent of the North Atlantic Ocean polygon in this data layer. Now, click OK to close the ENVIRONMENT SETTINGS window. Finally, click OK at the bottom of the POLYGON TO RASTER window to run the tool.

Right click on the name of your raster data layer (SBW_RANGE) in the TABLE OF CONTENTS window and select PROPERTIES. In the LAYER PROPERTIES window, click on the SYMBOLOGY TAB, select UNIQUE VALUES and then click OK to close this window. Next, in the TOOLBOX window, select SPATIAL ANALYST TOOLS> RECLASS> RECLASSIFY. In the RECLASSIFY window, select the SBW_RANGE data layer from the drop down menu in the INPUT RASTER section. Select VALUE in the RECLASS FIELD section. In the RECLASSIFICATION section of the RECLASSIFY window, type 1 in the first line of the NEW VALUES column. Next type 0 into the second line of the NEW VALUES column (which currently has the words 'NO DATA' in it). This will change all the No Data values for the data layer to zeros. Scroll down and make sure that there is no tick next to CHANGE MISSING VALUES TO NO DATA (OPTIONAL). Next, type C:\GIS_FOR_BIOLOGISTS\ SBW_RANGE_2 in the OUTPUT RASTER section, and then click OK. Finally, right click on the raster data layer named SBW_RANGE in the TABLE OF CONTENTS window and select REMOVE.

When the species range raster data layer is created, it covers a rectangular area defined by the extent of the NORTH_ATLANTIC_OCEAN_ALBERS data layer. However, this means that there are areas which are in the Pacific Ocean which are included in the resulting data layer. In order to restrict the coverage just to the area of interest (the North Atlantic) you will need to mask your species range data layer. In the case of this exercise, you will use the polygon representing the liimits of the North Atlantic Ocean in the NORTH_ATLANTIC_OCEAN_ALBERS data layer.

3. Mask the species range raster data layer so that it only covers areas in the North Atlantic Ocean

To do this, go to the TOOLBOX window and select SPATIAL ANALYST TOOLS> EXTRACTION> EXTRACT BY MASK. In the EXTRACT BY MASK tool window which will open, select SBW_RANGE_2 as the INPUT RASTER and select NORTH_ATLANTIC_OCEAN_ALBERS as the INPUT RASTER OR FEATURE MASK DATA. For OUTPUT RASTER, enter C:\GIS_FOR_BIOLOGISTS\SBW_RANGE_3 and then click on the OK button. Finally, right click on the raster data layer named SBW_RANGE_2 in the TABLE OF CONTENTS window and select REMOVE.

4. Set the symbols you wish to use to display your raster data layer in the MAP window

Right click on the name of your newly created raster data layer (SBW_RANGE_3) in the TABLE OF CONTENTS window and select PROPERTIES. **NOTE:** It may have been added below other data layers you already have in your data frame. Next, click on the SYMBOLOGY tab of the LAYER PROPERTIES window. In the left hand portion of the SYMBOLOGY window, select UNIQUE VALUES. Next, click on the ADD ALL VALUES button. Double click on the coloured rectangle beside the number 1, and select black for the colour. Do the same for the coloured rectangle next to zero, but select 10% grey. Finally, click the OK button. All the cells in the SBW_RANGE_3 data layer which fall in the range of Sowerby's beaked whale in the North Atlantic will now be coloured black, all other cells will be light grey.

284

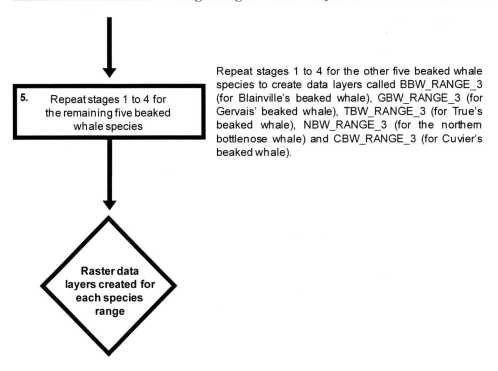

Repeat stages 1 to 4 for the other five beaked whale species to create data layers called BBW_RANGE_3 (for Blainville's beaked whale), GBW_RANGE_3 (for Gervais' beaked whale), TBW_RANGE_3 (for True's beaked whale), NBW_RANGE_3 (for the northern bottlenose whale) and CBW_RANGE_3 (for Cuvier's beaked whale).

At the end of this step, re-arrange the contents of your TABLE OF CONTENTS window until it looks like this (**NOTE:** The original range polygons for each beaked whale species will still be present in your GIS project, but will be positioned further down the TABLE OF CONTENTS window):

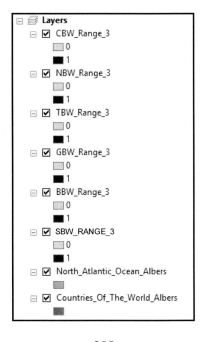

Now, click on the box next to all your newly created raster data layers in the TABLE OF CONTENTS window so that none of them are displayed in the MAP window. Repeat this for the original polygon data layers for each species range. Next, click on this box again for each individual raster data layer one at a time so that you can see the raster data layer of the range of each species. The raster data layer called SBW_RANGE_3 should look like this:

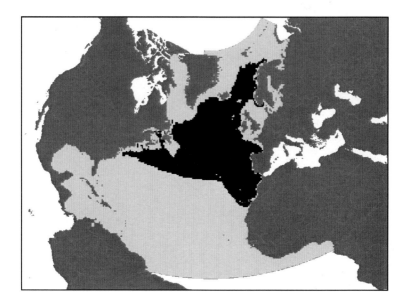

The raster data layer called BBW_RANGE_3 should look like this:

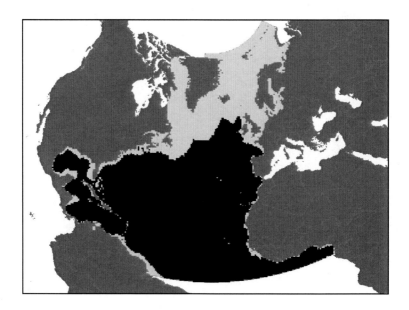

The raster data layer called GBW_RANGE_3 should look like this:

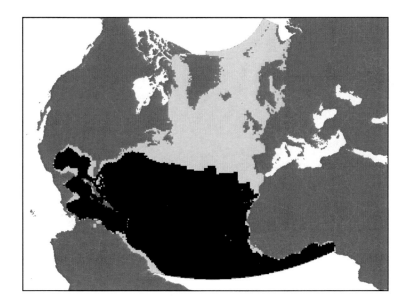

The raster data layer called TBW_RANGE_3 should look like this:

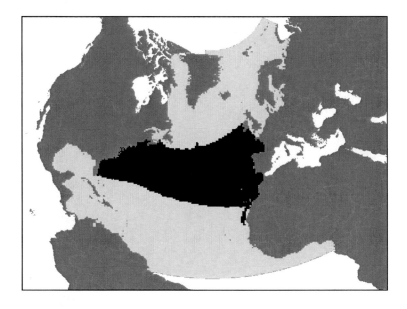

The raster data layer called NBW_RANGE_3 should look like this:

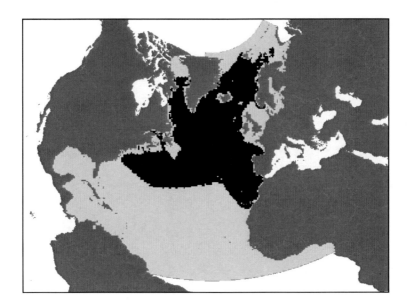

The raster data layer called CBW_RANGE_3 should look like this:

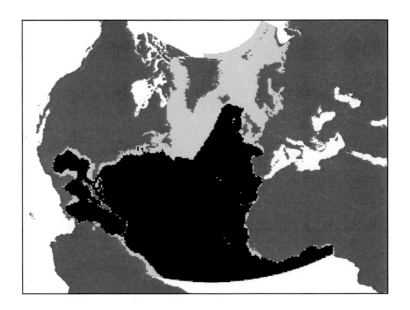

STEP 4: USE THE RASTER CALCULATOR TOOL TO CREATE A RASTER DATA LAYER OF BEAKED WHALE SPECIES RICHNESS:

Once you have created the raster data layers of the ranges of the individual species, you are ready to use the RASTER CALCULATOR tool to create a raster data layer of beaked whale species richness. This is simply the number of beaked whale species ranges which overlap in each 50km by 50 km grid cell. To create this species richness raster data layer, work through the following flow diagram:

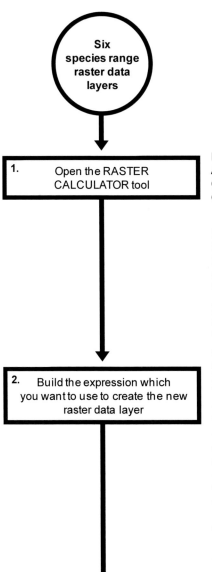

In the TOOLBOX window, select SPATIAL ANALYST TOOLS> MAP ALGEBRA> RASTER CALCULATOR. This will open the RASTER CALCULATOR window.

In the RASTER CALCULATOR window, you can build any expression you wish in the lower window. Existing raster data layers can be selected by double clicking on them in the LAYERS AND VARIABLES section of the window.

For this exercise, first double click on SBW_RANGE_3 to add it to the lower window. Then click on the button with the + sign on it. Next, click on BBW_RANGE_3, followed by the + button, and repeat this for GBW_RANGE_3, TBW_RANGE_3, NBW_RANGE_3 and CBW_RANGE_3 The final expression should read: "SBW_RANGE_3" + "BBW_RANGE_3" + "GBW_RANGE_3" + "TBW_RANGE_3" + "NBW_RANGE_3" + "CBW_RANGE_3". **NOTE:** The expression has to be entered exactly like this, including the spaces before and after the plus sign (+). This expression will produce a new raster data layer with values ranging from zero, where no beaked whales occur, to six, where the ranges of all six species overlap. Finally, in the OUTPUT RASTER section of the window, type C:\GIS_FOR_BIOLOGISTS\SP_RICHNESS and then click on the OK button to carry out this calculation.

289

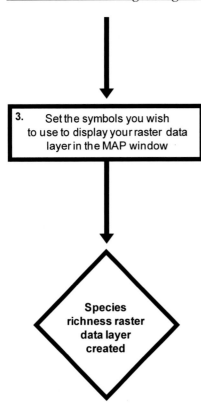

3. Set the symbols you wish to use to display your raster data layer in the MAP window

Species richness raster data layer created

Right click on the name of your newly created raster data layer (SP_RICHNESS) in the TABLE OF CONTENTS window and select PROPERTIES. **NOTE**: It may have been added below other data layers you already have in your data frame. Next, click on the SYMBOLOGY tab of the LAYER PROPERTIES window. In the left hand portion of the SYMBOLOGY window, select UNIQUE VALUES. Now, click on the ADD ALL VALUES button. Under COLOR SCHEME, select a colour ramp that grades from yellow at the left hand end to red on the right. Next, double click on the coloured rectangle beside the number 0 and select the second grey box down (which is 10% grey). Finally, click the OK button to close the LAYER PROPERTIES window.

At the end of this step, your TABLE OF CONTENTS window should look like this:

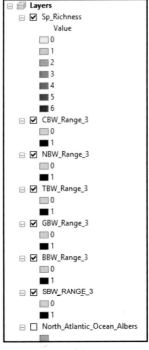

290

While the contents of your MAP window should look like this:

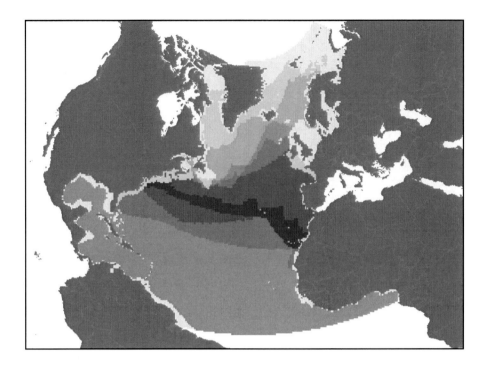

From the species richness raster data layer, you can see that the highest diversity of beaked whale species in the North Atlantic Ocean is found at mid-latitudes on the border between the warm temperate and sub-tropical realms, and it could be argued that conservation efforts to limit human impacts on beaked whales should be concentrated in this region. As you move out of this area, diversity decreases as you go both north and south. However, the reduction in species diversity with distance from this zone is greater to the north. In addition, within this raster data layer, you can see that beaked whale species diversity is also generally very low in coastal areas, such as the North Sea (the shallow area to the west of the UK) and along the continental shelf which lies off the coasts of the eastern United States and Canada. This is because the waters here are generally less than 200m in depth, and this is too shallow for the deep-diving beaked whales, which spend much of their lives at depths deeper than 500m. As a result, human activities which occur in such waters are unlikely to have a major impact on beaked whale species (although this does not mean that they will not affect other types of whales and dolphins).

This is the end of exercise six. In it you have learned how to create a species richness raster data layer from data layers which contain information about the ranges of individual species. In doing this, you have learned how to use the RATSER CALCULATOR tool to create new raster data layers by bringing together data from different individual raster data layers. This is a technique which is widely used when using GIS to answer biological research questions.

Instructions For QGIS 2.8.3 Users:

Once you have all the required files downloaded into the correct folder on your computer, and you understand what is contained within each file, you can move on to creating your GIS project for this exercise. The starting point for this is a blank GIS project. To create a blank GIS project, first open QGIS. Once it is open, click on the PROJECT menu and select SAVE AS. In the window which opens, save your GIS project as EXERCISE_SIX in the folder C:\GIS_FOR_BIOLOGISTS.

STEP 1: SET THE PROJECTION AND COORDINATE SYSTEM OF YOUR DATA FRAME:

As with all GIS projects, the first thing which you need to do when creating a GIS project to answer a specific research question is to work out what projection/coordinate system you are going to use. You are going to create a raster data layer of beaked whale species diversity which will cover the whole North Atlantic Ocean. This means that you need to use a projection/ coordinate system which will not cause any major biases over such a large area. One suitable projection is the Albers equal-area conic projection, and you will create a custom version of this projection centred on the middle of the North Atlantic. In addition, you will use the WGS 1984 datum for the coordinate system. To set your data frame to this custom projection/coordinate system, work through the flow diagram which starts on the next page. **NOTE:** In QGIS, the projection/coordinate systems are set using a short piece of code known as a Proj.4 string (see *http://github.com/Oseo/proj.4/wiki* for more details). It is very easy to enter these Proj.4 string codes incorrectly. As a result, the Proj.4 string code for the custom transverse projection/coordinate system which will be used in this exercise has been included in the compressed data folder for this book. It is called North_Atlantic_Projection_Proj4_String.txt, and it is recommended that you copy the code from this file and paste it into the appropriate place in the CUSTOM COORDINATE REFERENCE SYSTEM DEFINITION window (see the flow diagram which starts on the next page), rather than entering it yourself. This will ensure that you do not encounter any problems due to the use of the incorrect Proj.4 string to define this custom projection.

Data frame with no associated projection and coordinate system

When you open a new GIS project, there will be an empty data frame. It will not have a projection or coordinate system associated with it.

1. Create a new custom projection/coordinate system

To create a new custom projection/coordinate system, click on SETTINGS on the main menu bar and select CUSTOM CRS. In the CUSTOM COORDINATE REFERENCE SYSTEM DEFINITION window which opens, click on ADD NEW CRS button - it has a green plus (+) sign on it. Next, type NORTH ATLANTIC into the NAME section about half way down the page. This means that your new custom projection/coordinate system will be called NORTH ATLANTIC. Next, enter the following text into the PARAMETERS section (**NOTE**: To ensure this is entered correctly, you are advised to copy this code from the text file called North_Atlantic_Projection_Proj4_String.txt, which is in the folder C:\GIS_FOR_BIOLOGISTS, and paste it into the PARAMETERS section rather than trying to enter it manually yourself):

```
+proj=aea +lat_1=25 +lat_2=65
+lat_0=45 +lon_0=-30 +x_0=0 +y_0=0
+datum=WGS84 +units=m +no_defs
```

This is a PROJ.4 string which tells QGIS that your custom projection/coordinate system is an Albers equal area conic projection (aea) with the first standard parallel at 25 degrees north, the second at 65 degrees north, a latitude of origin of 45 degrees north and a central meridian of 30 degrees west based on the WGS 1984 datum (WGS84) and that the map units are in metres (m). Once you have entered this text, check it very carefully to ensure you have got it right and then click on the ADD NEW CRS button again. This will add it to the list of custom projection/coordinate systems in the top section of the window. Next, in this section, select NEW CRS and then click the REMOVE button – it has a red minus (-) sign on it. Finally, click OK to close the CUSTOM COORDINATE REFERENCE SYSTEM DEFINITION window.

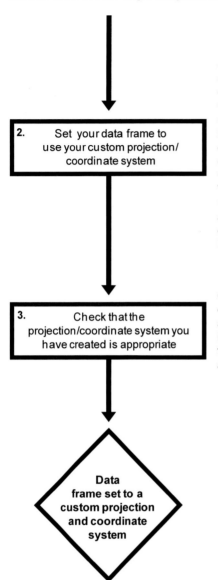

2. Set your data frame to use your custom projection/ coordinate system

In the main QGIS window, click on the CRS STATUS button in the bottom right hand corner (it is square with a dark circular design on it). This will open the PROJECT PROPERTIES CRS window. Click on the box next to ENABLE 'ON THE FLY' CRS TRANSFORMATION so that a cross appears in it. Next, type the name of the projection/ coordinate system into the FILTER section (in this case, it will be NORTH ATLANTIC). In the COORDINATE REFERENCE SYSTEMS OF THE WORLD section, click on NORTH ATLANTIC under USER DEFINED COORDINATE SYSTEM. This will add this to the SELECTED CRS section further down the window. Now click OK to close the PROJECT PROPERTIES CRS window.

3. Check that the projection/coordinate system you have created is appropriate

Once you have created a custom projection/ coordinate system, you need to check that it is appropriate. This involves examining how data layers look in it. For this exercise, this will be done in the next step by adding a polygon data layer of land and checking that it looks the right shape.

Data frame set to a custom projection and coordinate system

To check that you have done this step properly, click on PROJECT on the main menu bar and select PROJECT PROPERTIES. In the PROJECT PROPERTIES window which will open, click on CRS on the left hand side and make sure that the projection/coordinate system listed in the SELECTED CRS section is called NORTH ATLANTIC and that its definition in the window below contains the following text:

$$+proj=aea +lat\_1=25 +lat\_2=65 +lat\_0=45 +lon\_0=-30 +x\_0=0 +y\_0=0$$
$$+datum=WGS84 +units=m +no\_defs$$

If it does not, you will need to repeat this step to ensure that you have assigned the correct projection/coordinate system to your data frame. Now click OK to close the DATA FRAME PROPERTIES window.

STEP 2: ADD THE REQUIRED EXISTING DATA LAYERS TO YOUR GIS PROJECT:

Once the projection/coordinate system has been set for your data frame, you need to add the data layers which you will use for this exercise. These data layers are COUNTRIES_OF_THE_WORLD.SHP, NORTH_ATLANTIC_ OCEAN.SHP and the six polygon data layers which contain the information about the ranges of the six species of beaked whale regularly found in the North Atlantic. However, while all of these existing data layers use a coordinate system with the same datum as the one which you are using for this exercise (WGS 1984), they are in a different projection (the geographic projection rather than the custom Albers projection which will be used for this exercise). This means that they will need to be transformed into the required projection before they can be used in your GIS project. As was the case in exercise five, this will be done as part of the process of adding existing data layers to a GIS project. To add the required existing data layers to your GIS project, work through the following flow diagram:

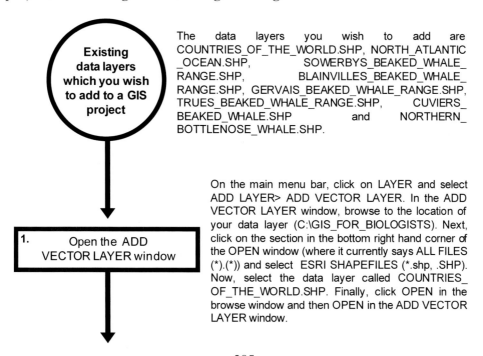

Existing data layers which you wish to add to a GIS project

The data layers you wish to add are COUNTRIES_OF_THE_WORLD.SHP, NORTH_ATLANTIC _OCEAN.SHP, SOWERBYS_BEAKED_WHALE_ RANGE.SHP, BLAINVILLES_BEAKED_WHALE_ RANGE.SHP, GERVAIS_BEAKED_WHALE_RANGE.SHP, TRUES_BEAKED_WHALE_RANGE.SHP, CUVIERS_ BEAKED_WHALE.SHP and NORTHERN_ BOTTLENOSE_WHALE.SHP.

1. Open the ADD VECTOR LAYER window

On the main menu bar, click on LAYER and select ADD LAYER> ADD VECTOR LAYER. In the ADD VECTOR LAYER window, browse to the location of your data layer (C:\GIS_FOR_BIOLOGISTS). Next, click on the section in the bottom right hand corner of the OPEN window (where it currently says ALL FILES (*).(*)) and select ESRI SHAPEFILES (*.shp, .SHP). Now, select the data layer called COUNTRIES_ OF_THE_WORLD.SHP. Finally, click OPEN in the browse window and then OPEN in the ADD VECTOR LAYER window.

2. Transform the data layer into the projection/coordinate system being used for your GIS project

In the TABLE OF CONTENTS window, right click on the name of the data layer which you have just added (COUNTRIES_OF_THE_WORLD) and select SAVE AS. In the SAVE VECTOR LAYER AS window, enter C:/GIS_FOR_BIOLOGISTS/COUNTRIES_OF_THE_WORLD_ ALBERS.SHP in the SAVE AS section and then select the option starting with PROJECT CRS from the drop down menu in the CRS section. Now tick the box next to ADD SAVED FILE TO MAP and then click on the OK button.

Once you have the transformed version of your data layer in your GIS project, you can remove the original version. This is done by right-clicking on its name (COUNTRIES_OF_THE_WORLD) in the TABLE OF CONTENTS window and selecting REMOVE.

3. Select how the data layer will be displayed in your GIS project

Right click on the name of the COUNTRIES_OF_THE_WORLD_ALBERS data layer in the TABLE OF CONTENTS window, and select PROPERTIES. In the LAYER PROPERTIES window, click on the STYLE tab on the left hand side. In the top left hand corner of the STYLE tab, select SINGLE SYMBOL and then click on FILL> SIMPLE FILL in the section below it. Now, on the right hand side of the window, click on the box next to COLORS FILL and select a dark green shade for the colour. Next, click OK to close the LAYER PROPERTIES window. You will see that the way the COUNTRIES_OF_THE_WORLD_ALBERS data layer is displayed in the MAP window has now changed to the new settings you have just selected.

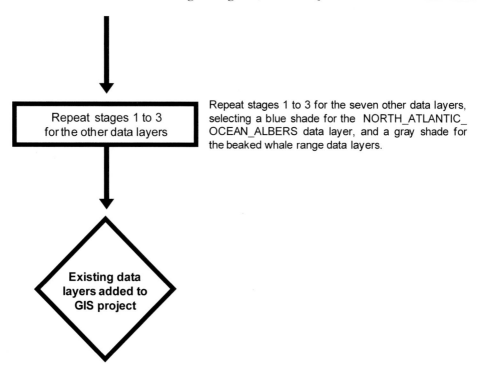

Repeat stages 1 to 3 for the other data layers

Repeat stages 1 to 3 for the seven other data layers, selecting a blue shade for the NORTH_ATLANTIC_ OCEAN_ALBERS data layer, and a gray shade for the beaked whale range data layers.

Existing data layers added to GIS project

At the end of this step, the TABLE OF CONTENTS window should look like this:

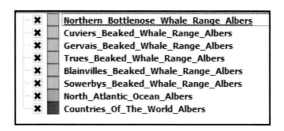

NOTE: If your data layers are not in this exact order, you can re-arrange them by selecting each one in turn, and while holding the left mouse button down, dragging it to its desired position.

Now right click on the data layer called NORTH_ATLANTIC_OCEAN_ ALBERS and select ZOOM TO LAYER. The contents of your MAP window should now look like this:

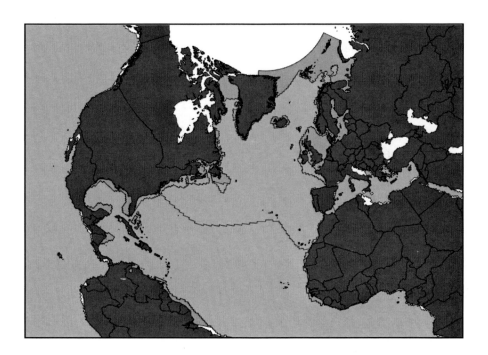

STEP 3: CREATE RASTER DATA LAYERS OF THE RANGES OF THE SIX NORTH ATLANTIC BEAKED WHALE SPECIES:

Once you have the required data layers added to your GIS project and transformed into the same projection/coordinate system as your data frame, you then need to convert each of the species range polygon data layers into a raster data layer. This is so that you can use the RASTER CALCULATOR tool to add up the number of beaked whale species which occur in each grid cell. Creating a raster data layer of species range from a polygon is relatively straight forward and can be done with the RASTERIZE conversion tool. However, once this data layer has been created, you need to do some further processing. This is in order to mask out any grid cells which fall in other ocean areas (such as the North Pacific). To make the species raster data layers, work through the flow diagram which starts on the next page.

Existing polygon data layer which you wish to convert into a raster data layer

1. Convert your polygon data layer to a raster data layer using the RASTERIZE tool

Click on RASTER on the main menu bar and select CONVERSION> RASTERIZE (VECTOR TO RASTER). This will open the RASTERIZE (VECTOR TO RASTER) window. For the INPUT FILE (SHAPEFILE) section, from the drop down menu, select SOWERBYS_BEAKED_WHALE_RANGE_ALBERS. Next, for ATTRIBUTE FIELD select EXPT_RANGE. Now, in the OUTPUT FILE FOR RASTERIZED VECTORS (RASTER) type C:/GIS_FOR_BIOLOGISTS/SBW_RANGE.TIF. Select RASTER RESOLUTION IN MAP UNITS PER PIXEL and enter a value of 50000 for HORIZONTAL and VERTICAL. This will set the cell size for the raster to a height and width of 50,000m (or 50km).

Next, you need to set the extent of the raster data layer which will be created. To do this, first click on the EDIT button (it has a picture of a pencil on it) at the right hand side of the section at the bottom of the tool window which has the code in it. Next, click in this section at the end of the existing code and press the space bar on your keyboard. Now type in the following code:

```
-te -6490106 -4874354 5029817 5472014
```

This code tells the software to create a raster with a bottom left corner at $-6,490,106$ and $-4,874,354$ and a top right hand corner at $5,029,817$ and $5,472,014$ (with the coordinates in the map units of the projection/coordinate system). Finally, click OK at the bottom of the RASTERIZE (VECTOR TO RASTER) window. Once the raster data layer has been created, click CLOSE in the RASTERIZE (VECTOR TO RASTER) window. **NOTE:** The coordinates used to set the extent of the raster data layer being created by this tool were selected as they represent the extent of the NORTH_ATLANTIC_OCEAN_ALBERS data layer and were obtained by right-clicking on the name of this data layer in the TABLE OF CONTENTS window and selecting PROPERTIES. In the LAYER PROPERTIES window, click on the METADATA tab on the left hand side and then scroll down to the PROPERTIES section. In this section, you can scroll down and find the coordinates which mark the extent of this data layer.

2. Mask the species range raster so that it only covers areas in the North Atlantic Ocean

When the species range raster data layer is created, it covers a rectangular area defined by the extent of the NORTH_ATLANTIC_OCEAN_ALBERS data layer. However, this means that there are areas which are in the Pacific Ocean that are included in the resulting data layer. In order to restrict the coverage just to the area of interest (the North Atlantic), you will need to mask your species range data layer. In the case of this exercise, you will use the polygon representing the limits of the North Atlantic Ocean in the NORTH_ATLANTIC_ OCEAN_ALBERS data layer.

To mask your species range data layer, click on RASTER on the main menu bar and select EXTRACTION> CLIPPER. In the CLIPPER window, select SBW_RANGE as the INPUT FILE (RASTER) and then enter C:/GIS_FOR_BIOLOGISTS/SBW_RANGE_2.TIF as the OUTPUT FILE. Next, tick the box next to NO DATA VALUE and select -1 from the drop down menu beside it. Now, in the CLIPPING MODE section, click on the MASK LAYER option and select NORTH_ATLANTIC_OCEAN_ ALBERS as the MASK LAYER. Make sure that the box beside LOAD INTO CANVAS WHEN FINISHED is ticked and then click on the OK button. Once the tool has finished running, click on the CLOSE button to close the CLIPPER window. Now, right click on the raster data layer you have just created (SBW_RANGE_2) in the TABLE OF CONTENTS window and select PROPERTIES. Once the LAYER PROPERTIES window has opened, click on the GENERAL tab on the left hand side (it has a picture of a hammer and a screwdriver on it), and click on the SELECT CRS button (it has a picture of a globe and a yellow box on it). In the COORDINATE REFERENCE SYSTEM SELECTOR window which opens, type NORTH ATLANTIC into the FILTER window, and select the custom NORTH ATLANTIC projection/coordinate system which is being used for this exercise. Now click on the OK button to close the COORDINATE REFERENCE SYSTEM SELECTOR window, and then click OK to close the LAYER PROPERTIES window. Finally, right click on the raster data layer named SBW_RANGE in the TABLE OF CONTENTS window and select REMOVE.

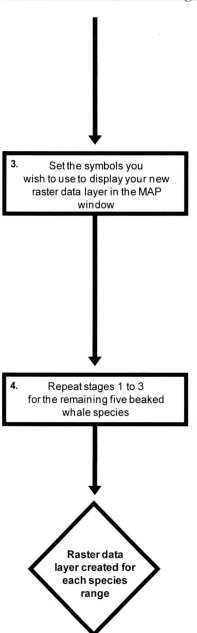

3. Set the symbols you wish to use to display your new raster data layer in the MAP window

Right click on the name of the SBW_RANGE_2 data layer in the TABLE OF CONTENTS window, and select PROPERTIES. In the LAYER PROPERTIES window, click on the STYLE tab on the left hand side. In the top left hand corner of the STYLE tab, select SINGLEBAND PSEUDOCOLOR from the drop down menu next to RENDER TYPE. Now, click on the plus sign (+) under COLOR INTERPOLATION to add a classification category to the table below it. In this table, enter 0 for VALUE and 0 for LABEL. Next, click on the box under COLOR and select a grey shade in the CHANGE COLOR window. Now, click on the plus sign (+) under COLOR INTERPOLATION again to add another classification category to the table below it. For this new category, enter 1 for value and 1 for LABEL. Next, click on the box on this row under COLOR and select black in the CHANGE COLOR window. Finally, click OK to close this window and then click OK to close the LAYER PROPERTIES window.

4. Repeat stages 1 to 3 for the remaining five beaked whale species

Repeat stages 1 to 3 for the other five beaked whale species to create data layers called BBW_RANGE_2 (for Blainville's beaked whale), GBW_RANGE_2 (for Gervais' beaked whale), TBW_RANGE_2 (for True's beaked whale), NBW_RANGE_2 (for the northern bottlenose whale) and CBW_RANGE_2 for (Cuvier's beaked whale).

Raster data layer created for each species range

Once you have finished this step, click on the box next to each of the polygon data layer representing a species range in the TABLE OF CONENTS window so that none of them are displayed in the MAP window. At the end of this step, your TABLE OF CONTENTS window should look like the image at the top of the next page.

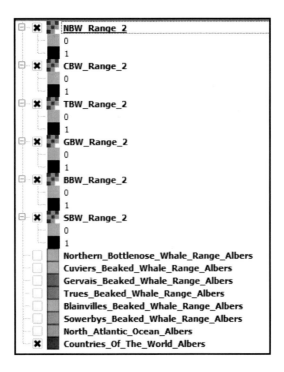

Now, click on the boxes next to all your newly created raster data layers in the TABLE OF CONENTS window so that none of them are displayed in the MAP window. Next, click on this box again for each individual raster data layer one at a time so that you can see the raster data layer of the range of each species. The raster data layer called SBW_RANGE_2 should look like this:

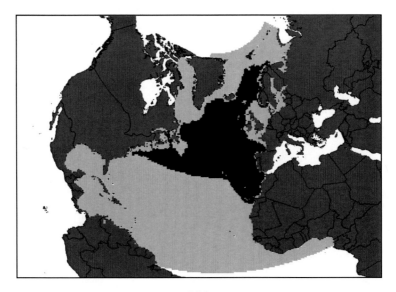

The raster data layer called BBW_RANGE_2 should look like this:

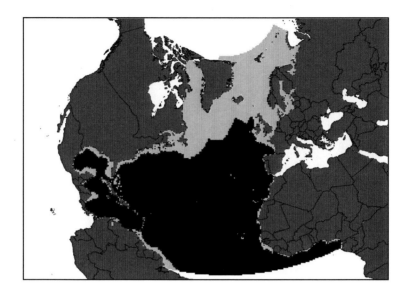

The raster data layer called GBW_RANGE_2 should look like this:

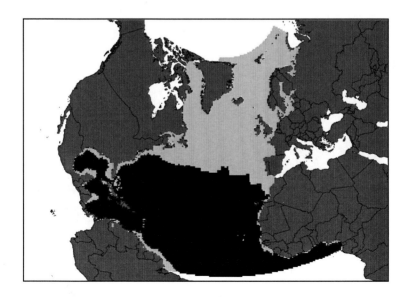

The raster data layer called TBW_RANGE_2 should look like this:

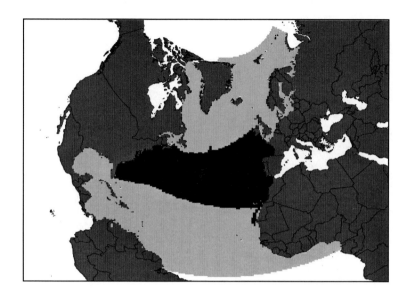

The raster data layer called NBW_RANGE_2 should look like this:

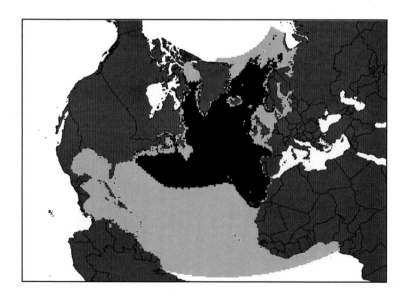

The raster data layer called CBW_RANGE_2 should look like this:

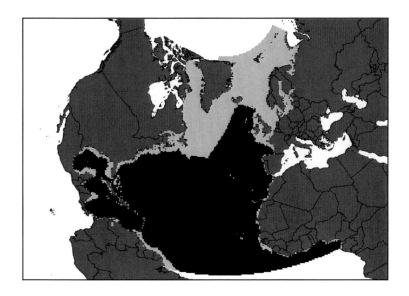

STEP 4: USE THE RASTER CALCULATOR TOOL TO CREATE A RASTER DATA LAYER OF BEAKED WHALE SPECIES RICHNESS:

Once you have created the raster data layers of the individual species, you are ready to use the RASTER CALCULATOR tool to create a raster data layer of beaked whale species richness. This is simply the number of beaked whale species ranges which overlap in each 50km by 50 km grid cell. To create this species richness raster data layer, work through the following flow diagram:

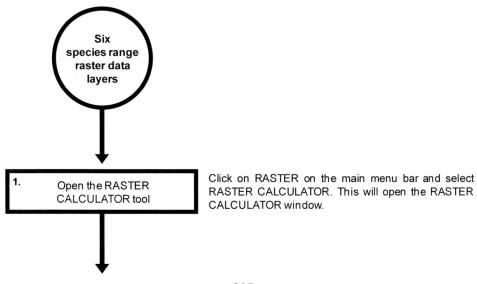

2. Build the expression which you want to use to create the new raster data layer

In the RASTER CALCULATOR window, you can build any expression you wish in the lower window. Existing raster data layers can be selected by double clicking on them in the RASTER BANDS section of the window.

For this exercise, first double click on "BBW_RANGE_2@1" to add it to the lower window. Then click on the button with the + sign on it. Next, double click on "CBW_RANGE_2@1", followed by the + button, and repeat this for "GBW_RANGE_2@1", "NBW_RANGE_2@1", "SBW_RANGE_2@1" and "TBW_RANGE_2@1". The final expression should read:

"BBW_RANGE_2@1" + "CBW_RANGE_2@1" + "GBW_RANGE_2@1" + "NBW_RANGE_2@1" + "SBW_RANGE_2@1" + "TBW_RANGE_2@1"

NOTE: The expression has to be entered exactly like this, including the spaces before and after the plus signs (+). This expression will produce a new raster data layer with values ranging from zero for cells which do not fall within the range of any beaked whale species, to six where the ranges of all six species overlap. Finally, in the OUTPUT RASTER section of the window, type C:/GIS_FOR_BIOLOGISTS/ SP_RICHNESS.TIF and click on the OK button to carry out this calculation.

3.   Set the symbols you
wish to use to display your
raster data layer in the MAP
window

Right click on the name of your newly created raster data layer (SP_RICHNESS) in the TABLE OF CONTENTS window and select PROPERTIES. **NOTE:** It may have been added below other data layers you already have in your data frame. Next, click on the STYLE tab of the LAYER PROPERTIES window. In the STYLE tab, select SINGLEBAND PSUEDOCOLOR from the drop down menu next to RENDER TYPE. Next, under GENERATE NEW COLOR MAP on the right hand side, select EQUAL INTERVAL from the drop down menu next to MODE, then enter 7 for CLASSES. Now, from the palette box above these, select the colour ramp called YlOrRd (this goes from yellow for the lowest values to red for the highest values) before clicking on the CLASSIFY buttton below it. This will add seven classes to the left hand section of the STYLE tab. In this section, double-click on the first row down under VALUE (where it currently says 0.000000 and enter the number 0. Repeat this for the same row under LABEL. Next, click on the next row down under VALUE and type 1. Repeat this for the same row under LABEL. Repeat this using the numbers 2, 3, 4 5 and 6 for each successive row. Now, double-click on the box under COLOR for the top row (which has a value of 0). This will open the CHANGE COLOR window. Select a grey shade and click OK. Finally, click the OK button to close the LAYER PROPERTIES window. Those cells in the SP_RICHNESS raster data layer which do not fall in the range of any beaked whale species will now be grey, while those which contain the range of one or more species will be coloured orange or red, and the redder the colour, the higher the species richness. All cells outside of the North Atlantic Ocean will be left blank and so will not be visible.

Species
richness raster
data layer
created

At the end of this step, your TABLE OF CONTENTS window should look like this:

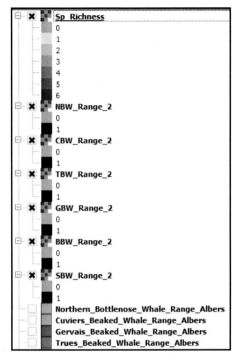

While the contents of your MAP window should look like this:

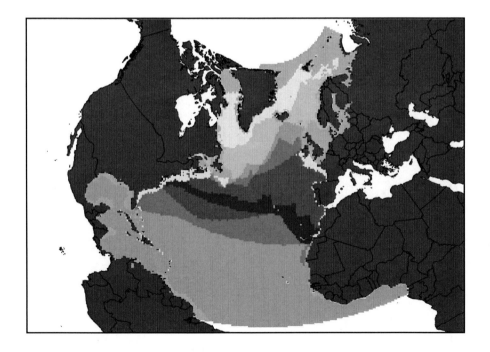

From the species richness raster data layer, you can see that the highest diversity of beaked whale species in the North Atlantic Ocean is found at mid-latitudes on the border between the warm temperate and sub-tropical realms, and it could be argued that conservation efforts to limit human impacts on beaked whales should be concentrated in this region. As you move out of this area, diversity decreases as you go both north and south. However, the reduction in species diversity with distance from this zone is greater to the north. In addition, within this raster data layer, you can see that beaked whale species diversity is also generally very low in coastal areas, such as the North Sea (the shallow area to the west of the UK) and along the continental shelf which lies off the coasts of the eastern United States and Canada. This is because the waters here are generally less than 200m in depth, and this is too shallow for the deep-diving beaked whales, which spend much of their lives at depths deeper than 500m. As a result, human activities which occur in such waters are unlikely to have a major impact on beaked whale species (although this does not mean that they will not affect other types of whales and dolphins).

This is the end of exercise six. In it you have learned how to create a species richness raster data layer from data layers which contain information about the ranges of individual species. In doing this, you have learned how to use the RASTER CALCULATOR tool to create new raster data layers by bringing together data from different individual raster data layers. This is a technique which is widely used when using GIS to answer biological research questions.

--- *Appendix I* ---

Trouble-Shooting GIS Projects

When things go wrong in a GIS project, it can be very difficult to work out what has happened and correct it. However, almost all problems boil down to just one basic issue which can be summed up by the question *Why don't my data layers appear where they should in my GIS project?*, or the more anguished variant *Where did all my data go?*

If this happens to you, the first thing to do is not panic and start randomly pressing buttons on your computer. In almost all cases, these problems are caused by one (or more) of the following thirteen basic issues:

1. You have zoomed too far in, too far out or panned away from the point in your data frame where your data are plotted. Try right-clicking on the name of the data layer in the TABLE OF CONTENTS window and selecting the ZOOM TO LAYER tool to take you back to the right place.

2. You do not have the right data frame active (this is only a problem in software packages, such as ArcGIS 10.3, where you can have multiple data frames in the same GIS project). Check whether you have the right data layer active. If not, activate the data frame which contains the data you wish to see displayed in the MAP window. In the ArcGIS 10.3 software package, this is done by right-clicking on the name of the data frame in the TABLE OF CONTENTS window and selecting ACTIVATE.

3. Your data layers do not overlay each other properly. They may plot in completely the wrong places, or be plotted much bigger, or much smaller than they should. Any such misplaced data layers can be located by using the ZOOM TO LAYER tools. This problem is usually caused when one, or more, of your data layers have not being assigned the correct projection (or coordinate system or datum), and, as a result, they are plotting in the wrong place. Alternatively, it may be that no projection has been assigned to one, or more, of your data layers. This can be solved by assigning the correct projection/coordinate system to each data layer where there is a problem.

4. When converting latitude and longitude to decimal degrees, you have applied the wrong formula (see appendix II for more information on the correct formula), or confused the format of your latitude and longitude data (for example, confusing data which are in degrees and decimal minutes for data which are in degrees, minutes and seconds or vice versa). In this case, your points will plot close to where they should, but not in exactly the right place, with some having a greater amount of error than others. To solve this problem, examine your original data, and do the conversion properly before loading the data into your GIS project again.

5. When converting data to decimal degrees, you have forgotten that latitudes south of the equator and longitudes west of the Greenwich Meridian should be negative and not positive (see appendix II for more details). This can result in points which should plot in the western Atlantic plotting in central Asia, or points which should be plotting off southern Africa, being plotted in Europe and so on. To solve this problem, go back to your original table and add the negative signs for southern latitudes and western longitudes and load the data into your GIS project again.

6. When creating a point data layer from data in a table, you have plotted latitude values as longitude values and longitude values as latitude values. Remember that longitude is the X coordinate and latitude is the Y coordinate, even though we usually say latitude first. Alternatively, you have used latitude for both the X and Y coordinates, or done the same with longitude, or you have accidentally selected the wrong field as the X and/or Y variable. This problem is solved by loading the data into your GIS project again, making sure that you use the correct fields for the X and Y variables.

7. You have some points which plot in the wrong place, such as in the sea off Africa, when they should be somewhere else, like central Asia. When you create a point data layer from a table, if any of the lines in your table are missing latitude and/or longitude values, these will be treated as having zero latitude and/or longitude and will be plotted in the according position. To solve this problem, remove any lines of data with missing latitude and/or longitude values from the table and load the data into your GIS project again.

8. You have moved the source files for a data layer and the GIS software no longer knows where to find it. This is usually indicated by the software telling you that the source file is no longer active, or that it cannot find it.

This can be solved in one of two ways. Either move the source files for the data layer back to their original locations, or add the data layer back in to your GIS project from its new location. A variation on this is that if you have stored files for data layers on removable disks, if these disks are not attached to your computer, the GIS software will not be able to find them.

9. You (or someone else) has edited a data layer you are using in your current GIS project in a different GIS project and, as a result, has changed the contents of the source file. Therefore, the features or information you are looking for in your data layer are no longer there. The only solution to this is to go back to the archived version of your data layer, make a copy of it, and load it back into your GIS project. If you have not created an archived version of your data layer, then you will need to try to find an original copy of it from somewhere else. However, any changes you have made since you last archived a copy of your data layer will be lost, so remember to archive your data layers regularly.

10. Last time you worked on your project, you forgot to save it before you closed it down. Unfortunately, the only way to deal with this issue is to add the missing data layers back into your GIS project. With any luck any changes you have made on your data layer will have been saved in them. If not, you will have to repeat the work. However, think of this as a learning experience and remember to save your project regularly as you go along, and definitely before you close it.

11. There is a conflict between your GIS software and your computer operating system in terms of the symbol it uses as a decimal separator. Specifically, if you are using a computer which is not set to use a decimal point (.) as the symbol to mark the division between whole numbers and decimal fractions (as in 12.345), this may cause problems. If you find you are having problems, you may need to change the settings on your computer to use a decimal point as the decimal separator. On computers running a Windows operating system, this can usually be changed by going to CONTROL PANEL> CLOCK, LANGUAGE AND REGION> CHANGE THE DATE, TIME, OR NUMBER FORMAT> ADDITIONAL SETTINGS> DECIMAL SYMBOL.

12. You have the wrong projection, coordinate system and/or datum assigned to one, or more, of your data layers. This is particularly a problem if you do not know what projection/coordinate system a data layer is based on and you try to guess it, or if someone has assigned the wrong

projection/coordinate system to one, or more, data layers. In addition, within the ArcGIS 10.3 software package, this problem can arise if you confuse the functions of the DEFINE PROJECTION tool with the PROJECT tool. The key difference here is that the DEFINE PROJECTION tool is used to assign the projection/coordinate system which a data layer was created in to that data layer (i.e. it creates a .PRJ file for that data layer). You cannot transform the projection/coordinate system of a data layer simply by changing the contents of the .PRJ file. Instead, to transform a data layer into a different projection/coordinate system, you need to use the PROJECT tool. To check the projection/coordinate system that has been assigned to a specific data layer, right click on the name of the data layer in the TABLE OF CONTENTS window and select PROPERTIES. The projection/ coordinate system (if one has been assigned to the data layer) will be listed in the DATA SOURCE tab (in ArcGIS 10.3) or the GENERAL tab (in QGIS) of the LAYER PROPERTIES window. If you have the wrong projection/coordinate system set for a data layer, you will need to change it to the correct one.

13. If you are having problems with raster data layers not plotting in the right place, this may because there has been a problem assigning the right projection/coordinate system to them. In particular, if you add a raster data layer to your GIS project and it has no projection/coordinate system assigned to it, you may find that even if you assign the correct projection/coordinate system to it, that it still does not plot properly. If this is the case, try removing the data layer from the GIS project, assign the correct projection/coordinate system to it and then add it back into your GIS project.

How To Convert Latitude And Longitude Values Into Decimal Degrees

In order to plot locational records in a GIS project, your latitude and longitude values need to be in decimal degrees. As long as you know the format which these values are in, it is relatively straight-forward to convert your data into decimal degrees. Details of how to do this for two common formats which latitude and longitude values are recorded in (degrees minutes and seconds, and degrees and decimal minutes) are provided below. However, if you do not know what format your latitude and longitude values are in, you cannot convert them to decimal degrees and so cannot use them in a GIS project. Therefore, it is essential that you record this information correctly and clearly.

Data in degrees, minutes and seconds should be recorded clearly as XX° XX' XX" with N after it for latitudes which are north of the equator or an S after it for latitudes which are south of the equator, and XXX° XX' XX" followed by E for longitudes which are east of the Greenwich Meridian or W for longitudes which are west of the Greenwich Meridian. These must include the unit markers which tell you that it is in degrees (°), minutes (') and seconds (").

Data in degrees and decimal minutes should be recorded as XX° XX.XXX' with N after it for latitudes which are north of the equator or an S after it for latitudes which are south of the equator, and XXX° XX.XXX' followed by E for longitudes which are east of the Greenwich Meridian or W for longitudes which are west of the Greenwich Meridian, using a decimal point to clearly separate out the minutes from the decimal fractions of a minute.

If you are working with a data set which contains records that you have not collected yourself, it may be difficult to tell what units have been used and, indeed, whether the same units have been used in every case. For example, in distributional data sets, it is quite common for some people to have submitted locational records in degrees, minutes and seconds and others in degrees and decimal minutes, but for all of them to have been entered as if they were in the same format. Such data sets are virtually useless for GIS analyses since it will be impossible to separate out which records have been recorded in which format. Therefore, if you are collating records from a wide range of sources pay particular attention to whether different people have used different units (despite any requests to use a specific format) when pulling the data together as this saves spending the time building a data set which you cannot analyse because you have positions recorded in mixed formats.

To convert latitude and longitude values to decimal degrees, work through the following flow diagram:

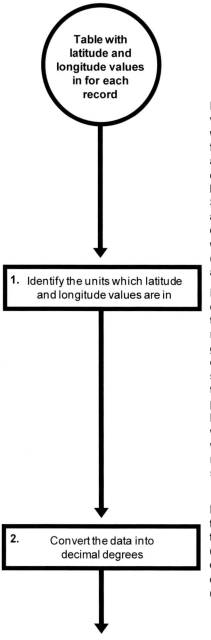

In order to be able to convert latitude and longitude values into decimal degrees, you must first know what units they are in. There are two common formats. These are degrees, minutes and seconds, and degrees and decimal minutes. In general, when data are in degrees, minutes and seconds, they will be written down as XX° XX' XX" with an N (north) or S (south) after it for latitude, and XXX° XX' XX" with an E (east) or W (west) after it for longitude. For degrees and decimal minutes, they will generally be written down as XX° XX.XXX' with an N (north) or S (south) after it for latitude, and XXX° XX.XXX' with an E (east) or W (west) after it for longitude.

If you are unsure which format some data are in, often the quickest way is to look at the last two or three digits (which will either be seconds or decimal minutes). If the lead digit for some of these is greater than 5, then it is most likely the data are in degrees and decimal minutes (since the value for seconds cannot go above 59). However, you need to make sure that you have this correct before proceeding. **NOTE:** This assumes that data have been recorded and entered in a consistent format, which is not always the case, and is something which needs to be ascertained and corrected, if required, before you do anything else with your data set.

If your data are in degrees, minutes and seconds, they can be converted into decimal degrees using the following formula: Degrees + (minutes/60) + (seconds/3600). If your data are in degrees and decimal minutes, they can be converted into decimal degrees using the following formula: Degrees + (minutes/60).

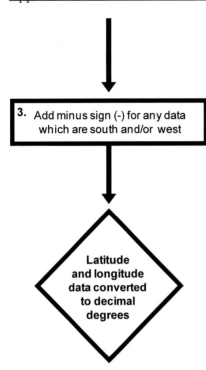

In decimal degrees, locations which are south of the equator are negative, while those north of it are positive. Similarly, locations which are west of the Greenwich Meridian are negative, while those to the east are positive. It is best to add any required minus signs after converting to decimal degrees as adding a negative sign can affect the calculation of the formula if it is not added correctly.

A spreadsheet containing the formulae for converting latitude and longitude values into decimal degrees can be downloaded from the *Useful Tools* page on the *GIS In Ecology* website (*tinyurl.com/GFB-Link9*).

--- *Appendix III* ---

Glossary Of Common Terms In GIS

Accuracy: This is how close a value in a data set is to the true world value that it is meant to represent. In GIS, there are two areas where accuracy is important. The first is how accurately values within data layers correspond to real world values they are meant to represent. For example, within a raster data layer of land elevation an individual grid cell value may be 200m. If the true altitude for that particular location is 205m, then the elevation grid could be said to have a high level of accuracy. If, however, the true altitude is 20m, then the elevation grid could be said to have a low level of accuracy. The second area where accuracy is important is the spatial location of features within a data layer. A high level of spatial accuracy is important to ensure that the spatial relationships between features within a GIS are accurate. In general, you should try to ensure that when you are recording positions of features, such as study sites or locations of species records, you do it with the highest level of accuracy possible. The use of GPS receivers has greatly increased the ability to accurately record locations (down to an accuracy of ~15 metres). However, this does not mean that GPS positions will always be accurate, and problems with receiving signals from multiple satellites can mean that sometimes GPS positions will not be as accurate as they should be, even if they are provided on the GPS receiver in a precise manner (see *www.GISinEcology.com/GFB.htm #video3* for an illustration of the difference between precision and accuracy).

Assign Data By Location: Another name for a Spatial Join. See Spatial Join for more information.

Attribute: This is information about a single piece of data (or feature) in a feature data layer. Examples of attributes might include information about its position (e.g. its latitude and longitude), what was recorded there or when it was recorded. Generally, these are displayed in a table (the attribute table) connected to a feature data layer in which each attribute represents a single field or column. If lots of single pieces of data have the same attributes, they can be grouped in the same data layer because they have the same fields. This, of course, does not mean that they have to be. For example, even though you have the same information about every sighting of a particular taxonomic

317

group, you might want to have data on the locations where each individual species was recorded in a separate data layer. If, however, individual features, or groups of features, have different attributes it may be difficult to group them into a single data layer because the fields in the attributes table will not match up.

Attribute Table: A table that accompanies a feature data layer, such as those containing points, lines and polygons, which contains information on the attributes of each feature within that data layer.

Buffer: An area within a specified distance of a feature or features in a data layer. For example, a buffer may be used to identify the areas surveyed on either side of a survey track (also known as the survey swath). In general, buffers can be based either on a fixed distance or a variable distance based on the values in a field of the attribute table for the data layer of interest.

Cell Size: A cell represents the smallest part of a grid or raster data layer. Each cell within a grid is associated with its own value. The size of the cell will define the resolution of that particular grid or raster data layer.

Clip: To use the features of one data layer to remove parts of another data layer based on their spatial overlay. As such, 'Clipping' can be thought of as very much analogous to Masking, although they may be given different names within GIS software.

Continuous Surface: A continuous surface is a data layer which covers the entire area within its extent. Therefore, each point in space is represented by one (and only one) value in the data layer. This may also include a value which indicates when no data are present for a specific location. Continuous surfaces may be non-gridded continuous surfaces, such as triangulated irregular networks (TINs), gridded continuous surfaces, such as raster data layers, or, in the case of polygon grids, feature data layers.

Coordinate Reference System (CRS): The coordinate reference system, or CRS, is an alternative name for the projection/coordinate system of a GIS project or data layer.

Coordinate System: Coordinates are the address for any individual point in a GIS. Typically, these are in the form of an X coordinate, or easting, which tells you how far left or right a point is from a given starting point or origin, and a

Y coordinate, or northing, which tells you how far up or down it is from this same point. The units (known as map units) for the coordinate system are usually defined by the projection it is associated with. For example, for a transverse mercator projection, the coordinates are usually given in metres, while for the geographic projection, the units are decimal degrees (representing longitude and latitude). When written down, the value for the Y axis (the northing or latitude) may sometimes be given first, followed by the X coordinate (the easting or longitude). Further considerations of the issue of coordinate systems can be found in chapter four.

Data Frame: A data frame represents the section of the Earth which is being covered by a GIS project. It can be thought of as being analogous to an interactive map and it represents the space where all the data layers within a GIS project are grouped in such a way that they can all be displayed on top of each other and can interact with each other. **NOTE:** There may be problems using data layers in the same data frame if they have different projections, coordinate systems or use different datums to each other and/or from that used for the data frame itself. Therefore, it is good practice for novice GIS users to always ensure that all data layers are in the same projection/coordinate system as the data frame they are being used in.

Data Layer: A data layer is a data set which contains a specific set of information in a specific way in a GIS project. For example, it might represent elevation data as contour lines, or sampling sites as points. You cannot have data in more than one format in the same data layer. For example, you cannot have elevation represented as contour lines and points in the same data layer, as you might see on a terrestrial map or nautical chart. More information about the different ways data can be represented in individual data layers can be found in chapter five. Data layers are often stored as shapefiles which consist of a number of different files and/or folders with the same name, but with different suffixes such as .shp and .dbf. All these separate files are needed for a data layer to be used in a GIS project. As a result, if you are copying any data layer files outside of your GIS software, always make sure that you copy all parts of it.

Datum: A datum is the starting point on a specific spheroid from which all other points are measured and, therefore, which defines the starting point for the coordinate system. As you can imagine, if different data sets use different datums, then the same point will be plotted in different places (which is not good). As a result, you need to know what datum your data set is based on. Luckily, almost all modern coordinate systems now use the WGS 1984 datum.

This is the one used by Global Positioning System (GPS) satellites and most remote sensing satellites. However, if you are using a GPS receiver, these can be set to different datums, so check that you know how yours is set. Similarly, older paper maps, charts and data sets may use different datums. For any one coordinate system, using the wrong datum can result in an error of up to 1km or more. As a result, this is not so important for broad scale and coarse resolution studies, but it is critical for local scale and fine resolution ones. Further considerations of the issue of datums can be found in chapter four.

Decimal Degrees: Decimal degrees are a unit of latitude and longitude. They are generally in the format 12.345678, and the number of decimal places in decimal degrees represents the resolution of such data. The figures after the decimal point represent the minutes, or minutes and seconds, of latitude or longitude as a fraction of a degree. There are 60 minutes in a degree and 60 seconds in a minute. In decimal degrees, coordinates which are north of the equator (0 degrees latitude) and east of the Greenwich Meridian (0 degrees longitude) are positive, while those which are south and west of these lines are negative. Latitude and longitude data must be converted into decimal degrees in order to be plotted in a GIS. It is important when doing this conversion that you know whether your original data are in degrees, minutes and seconds (the traditional format) or degrees and decimal minutes (the standard format for most GPS receivers). More information on converting positions into decimal degrees can be found in Appendix II on page 314.

DEM: See Digital Elevation Model.

Digital Elevation Model: A digital elevation model, or DEM for short, is a representation of the topography of the Earth's surface. In biology, this generally refers to the topography of the land or seabed. DEMs can be represented in a number of ways in a GIS, but typically are either a raster grid or a triangulated irregular network (TIN) – see chapter five for more details.

Digital Number: A digital number is the name given to the value of a pixel in a satellite or remote sensing image which does not represent the true value of the environmental variable it represents. Rather, it represents this value on a different scale (usually an ordinal scale between 0 and 255). In order to use the true value of such images in a GIS project, you will need to convert the digital number using the appropriate conversion equation. This can be done by turning the image into a raster data layer and then applying the conversion equation using Map Algebra.

Euclidean Distance: A Euclidean distance is a distance measured assuming that two points are on the same flat surface. Thus, a Euclidean distance between two points as measured in a GIS may vary depending on the projection, coordinate system and datum used to represent these points on a flat surface. As a result, the Euclidean distance measured in a GIS may differ from the actual distance between the two points in the real world if an inappropriate projection, coordinate system or datum is used to display the points on the flat surface of a GIS map. This represents one of the main reasons why it is essential to use the correct projection for any GIS project. See chapter four for more information.

Extension: An extension is an add-on tool that can be used to extend the functions of GIS software. In many cases, these may have been written by people other than the original software developer, and are often key to using GIS for biological research as they may provide tools for doing tasks which, while not common in terrestrial geography, are frequently done by marine biologists. For example, the Hawths Tools extension for ArcView 9.3 and Geospatial Modelling Environment for ArcGIS 10 are key for calculating the home ranges of individual animals from locations where they have been recorded. In general, extensions must be loaded in and activated in order to use their functions. Useful extensions can usually be found by relatively simple internet searches.

Extent: Extent refers to the area covered by an individual data layer or data frame. For raster data layers, it is set by the coordinates which define the corners of the grid, or by the number of rows and columns, and cell size of the grid.

Extract: To take a subset of the information from a data layer and use it to make another data layer (i.e. the data are extracted from a specific data layer) or to add a subset of information to the attribute table of another existing data layer (i.e. the data are extracted to a second data layer). This may involve taking only data from a certain spatial area, data with a specific range of attribute values, or taking data from one data layer and adding it into a second different data layer. For example, a common task in biology is to extract elevation data from one data layer and add it into a data layer of records for a particular species based on the locations of those records. This would then allow you to analyse the habitat preferences for that species.

Feature Data or **Feature Data Layer:** Feature data layers are data layers which contain points, lines or polygons (known as features), each of which are defined by a set of X and Y coordinates (also know as vectors) that mark where they should be plotted in a GIS. Feature data layers are usually accompanied by an attribute table, which is essentially a list that contains information about each point, line or polygon within that specific data layer. A feature data layer cannot contain multiple types of features. For example, a feature data layer can contain either points, lines or polygons, but it cannot contain points and lines, lines and polygons, points and polygons, or all three types of feature together. Since feature data layers are essentially a list of coordinates which are grouped to define individual features, they are not necessarily dependent on the projection and coordinate system in which they were created, and they can easily be transformed into different projections and coordinate systems. While the relative positions of points and lines, and the shapes of polygons, may appear to change during such a transformation, they will still represent the same locations on the surface of the Earth. A feature data layer can represent a grid, where each grid cell is a separate polygon, with accompanying information in an attribute table. This is known as a polygon grid data layer to separate it from a raster grid data layer. An example of how a polygon grid data layer can be used in biological research can be found in the exercise outlined in chapter fifteen.

Feature: A feature is a single element within a feature data layer (that is, one containing points, lines or polygons). A data layer may consist of a single feature or it may consist of a large number of different features. Each feature is represented by a unique line in the attribute table and represents the smallest selectable element within a data layer. In order to select only part of a feature, it must first be divided in some way into two separate features represented by their own individual lines within the attribute table of the data layer. The type of feature (that is, whether it is a point, a line or a polygon) within a data layer will depend on how the data are represented within a GIS project.

Field: A column within the attribute table of a data layer containing information about a specific attribute.

Geodatabase: A geodatabase is a data management and storage framework which is specific to ESRI's ArcGIS software. It allows all data to be stored in a single location and also allows access to some of the more specialist GIS tools which come with that software package. However, for beginners, it is often easier to use what is known as the 'shapefile' approach for storing the data for their GIS projects. In the shapefile approach, which is used for the exercises in this book, each data layer is stored as an individual shapefile within a specific folder (usually on the C-drive of a computer). This allows for easier access to individual files, easier transfer of data layers to other GIS software packages and avoids beginners having to learn about the structure and functioning of geodatabases before they can start using GIS in their research. More experienced users, however, may prefer to use the geodatabase approach to storing data (if their chosen GIS software package allows it). Regardless of the approach used, the same basic rules apply when using GIS for biological research.

GPS or Global Positioning System: This is a network of satellites which transmit radio signals. When received by a GPS receiver, these signals can be used to calculate the latitude and longitude of the receiver. While this position will also be that of the observer for hand-held GPS units and in small boats, on large ships, the position will represent the position of the receiver's antenna, which may be some distance from where the observer is positioned. While this is generally not an issue, it may be under some circumstances and is something you should be aware of.

Graphic: A graphic is an element which is present in a GIS project but which is not part of a data layer. For example, labels providing information about individual features can often be added as graphics in a data frame. These can be edited or moved around without the need to edit the contents of a specific data layer.

Graticules: Graticules are the markers which provide information about the latitude and longitude for a map. These may be included as numbers around the edges of the map, or as markers on the map itself.

Grid: A grid is a set of regularly-spaced data and is one of the most common ways to divide up a study area into different locations for sampling or other purposes. In addition, it is a common way of dividing a continuous set of data (e.g. land elevation) into discrete values associated with individual locations. Grids can be displayed in a number of different ways within a GIS project (see chapter five for more information).

Interpolation: Interpolation allows you to fill in areas between neighbouring data points of known values. For example, interpolation can be used to convert elevation data points into a continuous surface or grid (an example of this can be found in exercise three in chapter fourteen). Interpolation can be done in a number of different ways (such as inverse distance weighting or kriging) and can be based on varying numbers of neighbouring points. Which approach and what number of points are used will vary depending on exactly what you wish to do. For elevation or water depth data, using inverse distance weighting based on three neighbouring points is typically a useful starting point.

Intersect: To use features from one data layer to cut the features in another data layer into smaller pieces based on how they overlay each other. For example, if you have a polygon data layer representing different parts of your study area, and you wish to work out how much survey effort you conducted in each part, you can use an intersect tool to cut the survey lines (which will be in a line data layer) into separate lines for each part of the study area (see exercise four in chapter fifteen). In addition, it will usually add information to the line data layer identifying which part of the study area each section of the lines are in.

Join: A join is when two or more data sets are combined based on a common attribute in a shared field of a table. For example, an identification code could be used to join data from different attribute tables if the same identification codes apply to the same feature in each table.

Layout: This is a way to view the data in your GIS project which allows you to create a figure or map based on its contents. In general, a layout will be attached to a specific data frame, and only those elements displayed in that data frame will be shown in it.

Legend: A legend provides information about the way the data within a data layer are displayed in a GIS project. For example, all features within a data layer with the same value for a specific attribute may be displayed in the same colour, while all features with different values for that attribute are displayed in other colours. The legend tells you what each colour represents. Changing the legend only changes the way in which the data are displayed and does not change the contents of a data layer, or anything in its attribute table.

Locational Data: These are data which refer to an individual point in space, and generally represent a point where something happened (such as a sampling station) or was observed (such as a sighting of a particular species of interest). They are generally recorded as a set of coordinates along with additional non-spatial information such as time, date, what was recorded and who recorded it. As such, they usually represent 'X-Y Data' and can be treated the same as any other X-Y data set. However, on some occasions, locational data will be recorded as 'waypoints' on a GPS receiver and can be entered into a GIS more directly (see chapters nine and thirteen for more information).

Map Algebra: Map algebra provides a way of carrying out calculations based on raster data layers. This can involve either a single raster data layer (such as when converting the digital numbers in a satellite image for sea surface temperature or chlorophyll into the real values which they represent), or a number of raster data layers (such as when making visualisations of the predicted occurrence of a species from a habitat preference model). Map algebra works by conducting the required calculation separately on every cell within a raster data layer to produce a new data layer with the same extent and cell size with the results of the calculation as the new value for each cell. When using multiple raster data layers in a single calculation, it is generally useful to ensure that they all directly overlay each other and have the same extent. In GIS software packages, map algebra is generally conducted using a raster calculator tool (see exercise six in chapter seventeen for an example of using a raster calculator tool to do map algebra in a biological context).

Map Units: The map units are the units in which the coordinates are provided for a coordinate system. This means these are the units in which distances and areas are measured for a specific projection. In many cases, these will be metres (as is the case with a transverse mercator projection), but this is not always the case. For instance, in the geographic projection, the map unit is decimal degrees. This is important because only when the map unit is in a real measure of distance (such as metres) can distances, areas and other measures be properly calculated. For example, the slope of a hill represents the change in elevation between two points divided by the distance between the points. As a result, slope cannot be calculated when the map unit is not a true measure of distance. This is a common problem with using the geographic projection, and this is the reason its use should be avoided in most biological GIS projects.

Mask: A mask is a data layer which is designed to define areas that need to be removed from a specific data set. For example, in marine biology it is common to use a mask to remove any points which fall on land. However, masks can also be used to exclude any data points which fall outside of a study area, which fall more than a specific distance from a feature of interest, or which fall outside a specific altitude. For example, a mask based on the position of the 200 metre elevation contour can be used to exclude all points that do not fall in lowland areas. Masks may be based on other existing data layers or they may be created from scratch.

Masking: To use a mask to exclude data or features from specific areas within a data layer.

Merge: To combine separate features within a data layer into single features based on the values of a selected attribute.

Neighbourhood Statistics: Neighbourhood statistics allow you to compare individual cells within a raster data layer to those which surround it. This is done by defining how many grid cells are to be compared and in what way. For example, neighbourhood statistics can be based on a square of cells that surround individual cells, or on a given radius around individual cells. This defines the 'neighbourhood' used to calculate the statistics. The statistics which can be calculated include a mean value, a standard deviation, a majority value or the variability in the 'neighbourhood'.

'No Data' Value: A 'no data' value is a value for a cell in a raster data layer which indicates that it contains no information. This is different to a cell containing a value of zero. This is required because all cells within a raster data layer must have a value and cannot be left blank. By including a 'no data' value, GIS software can be told that a specific cell contains no real information and should be treated as if it was empty. Typically, 'no data' values are selected based on the requirements of a specific programme, or on the requirements of a specific data set. The only requirement is that it must not be in the range of real values in the raster data layer. However, in order for a raster data layer to function properly, you may need to tell your GIS software what this value is. If you do not set the 'no data' value, it may not display properly or may cause problems when you use the data layer.

Overlay: When one data layer sits spatially on top of part or all of another data layer they are said to overlay each other. Two data sets must overlay each other if information needs to be extracted from one based on the positions of features in the other. In many cases, when working with multiple raster data layers, it is essential that the raster data layers overlay each other exactly. That is, their extents must be exactly the same and the edges of their cells must also line up with each other. Data layers which use different projections, coordinate systems and/or datums may need to be transformed into the same projection, coordinate system and datum in order to ensure they overlay each other properly.

Polygon Grid Data Layer: This is a polygon data layer where each polygon represents a single, non-overlapping cell in a grid, with information stored about each polygon grid cell, such as a unique ID number, in an accompanying attribute table. In this attribute table, each polygon grid cell is represented by a single record or line, with the values for each attribute provided in each field or column. Polygon grids should not be confused with raster grids, even though both store information about the values for specific attributes for individual grid cells, and can cover the same areas in space. In particular, they are used for different purposes in GIS in biological research (see exercises three and four in chapters fourteen and fifteen, respectively).

Precision: This is how precisely a value is measured or expressed. For example, elevation expressed to the nearest 100m will be less precise than elevation expressed to the nearest 1m. However, just because a data layer has a high level of precision, this does not necessarily mean that it has a high level of accuracy. For example, elevation may be measured to the nearest metre, giving it a high level of precision. However, if there is a problem with the altimeter being used to measure elevation above sea level, meaning that all measurements are off by 50 metres (for example, due to poor calibration), the elevation measurements will not be accurate, no matter how precise they are (see *www.GISinEcology.com/GFB.htm#video3* for an illustration of the difference between precision and accuracy).

Projection: A projection is a way of displaying the surface of a three dimensional object (in this case planet Earth) on a two dimensional flat surface, such as a piece of paper (in the case of a traditional map or chart) or a computer screen (in the case of GIS software). As you might imagine, this is not a simple process and always leads to some distortion. As a result, there are many different types of projections, each of which distorts the Earth's surface in a different way to make it fit onto a flat surface. In general, for GIS projects in biology, the most appropriate projection is likely to be either a local, regional or nation-specific projection, such as the British National Grid projection. If one is not available, a transverse mercator (for study areas <~500km in radius) or Lambert azimuthal equal area (for study areas <~1,000km in radius) centred on, or close to, the middle of the study area will usually suffice. However, once you move a long way away from the centre of such projections, the relationships between points start to become distorted. As a result, for larger study areas (such as whole regions or continents), other projections may be more appropriate depending on exactly what you wish to do with your GIS project. For example, you may need a different projection to accurately measure large areas than to accurately measure long distances. In addition, different projections may be required for long (>1,000km), narrow study areas than for square ones in order to minimise the distortion along its entire length. A brief video explaining what projections are can be found at *www.GISinEcology.com/GFB.htm#video3*, while advice about picking the right projection from those available is covered in more detail in chapter four.

At this point, it is worth mentioning the so-called geographic 'projection'. This is not a true projection and it uses decimal degrees as its unit of measurement. This means that it can be used to plot the entire globe. However, a decimal degree does not represent the same distance in all parts of the world. For example, a decimal degree of latitude is shorter, in real terms, at the equator than at the poles due to the curvature of the Earth. In addition, a decimal degree of longitude does not necessarily equal a decimal degree in latitude in terms of real distances. As a result, the geographic projection should not be used for measuring distances or doing calculations based on distances (such as calculating slope). Projections are one of the most important things to understand in GIS and using the wrong projection or trying to compare data which use different projections without taking this into account are some of the most common mistakes made when using GIS.

Query: A query is a way of selecting part of a data layer based on the values in its attribute table (sometimes referred to as a table query) or its spatial location relative to features in another data layer (sometimes known as a spatial query).

Raster Calculation: An alternative name for Map Algebra.

Raster Data, Raster Grid or **Raster Data Layer:** A raster data layer is a data layer which is arranged in a grid, or an array, format. However, unlike feature data layers, each cell within this grid is not a separate feature, and raster data layers are not accompanied by an attribute table which lists the information about each individual grid cell. Instead, the data are stored as an array of rows and columns which give the value for each grid cell, along with information on the size of each grid cell (the height and the width), the number of grid cells in each row and column, and the position of the centre of one of the grid cells (usually the lower left hand grid cell). This information allows the GIS software to know where the raster data layer should be plotted. In general, each cell in a raster data layer is square, and, thus, a raster data layer is very much dependent on the projection and coordinate system which it was created in. As a result, it can be difficult to transform a raster data layer into a different projection and coordinate system. Raster data layers are also known as raster grids to separate them from polygon grids.

Reclassify: Reclassification changes the existing values in a raster data layer to new values based on the original values for each cell. Reclassification can be used to change a continuous variable into an ordinal or nominal scale. For example, reclassification can be used to change the values in an elevation raster data layer from a continuous scale into one which separates a study area into two habitat categories (lowland and highland) based on a specific elevation contour which marks the divide between the two habitat types.

Resample: Resampling allows you to change the resolution or size of cells of a raster data layer. For example, it allows you to divide up a raster data layer with a resolution of 5km by 5km to one with a resolution of 1km by 1km (making it a finer resolution grid) or 10km by 10km (making it a coarser resolution grid).

Resolution: The resolution of a data layer has three related meanings, one in relation to line and polygon data layers, one relating to point data layers and another relating to raster data layers. In line and polygon data layers, the resolution is the smallest features or parts of a feature which are, or can be, represented. For example, a line data layer representing elevation contours would have a resolution of 250m if the smallest topographic features represented within it are those which are 250m, or more, across. In a point data layer, the resolution is the shortest distance over which two events occurring in quick succession would be considered separate points and so be represented by separate rows within the attribute table of the data layer. For example, in a survey of whales, it would represent the minimum distance at which two groups of whales, seen in quick succession, would be entered into a point data layer as two different points rather than one point where two groups were recorded. In a raster data layer, resolution represents the size of grid cells within it. A fine resolution grid will have a small grid cell size while a coarse resolution grid will have a large grid cell size. Within any particular study, you may use raster data layers with different resolutions. For example, you might use a fine resolution raster data layer to look at distribution in relation to elevation, but a coarse resolution raster data layer to look at the distribution in relation to climatic variables, such as rainfall or average summer temperature.

Scale: Rather confusingly, scale is used to refer to two different concepts in GIS. Firstly, scale is used to refer to the resolution used in an analysis or a data layer. For example, a fine scale study of habitat preferences of dolphins might use a depth grid with cells that are 500m by 500m, while a coarse scale study of habitat preferences might use a depth grid of 10km by 10km. Secondly, scale is used to refer to the size of the study area. For example, a local scale study might look at habitat preferences of dolphins within a study area that is a few tens of kilometres across, while a broad scale study might look at habitat preferences across hundreds to thousands of kilometres. This book will use resolution for the first meaning, and will only use scale to refer to the size of the study area.

Select By Data Layer: To use the features from one data layer to select features in another data layer. Features can be selected in a number of possible ways depending on how they overlay each other. For example, features can be selected based on whether their centres fall within a feature of another data layer, or whether all of it is contained within a feature of another data layer. 'Select By Data Layer' can be thought of as a spatially-explicit query which can be used to select data based on their spatial relationships with other data layers rather than by values in the attributes table, and, as a result, is sometimes known as a Spatial Query.

Shapefile: A shapefile is a specific, and widely used, format for feature data layers, such as points, lines or polygons.

Spatial Join: A spatial join is when two, or more, data sets are joined based on their spatial locations rather than on common attributes in a shared field. For example, the data from an elevation data layer may be joined to a data layer containing the locations where a particular species has been recorded based on their positions. The ability to do spatial joins is one of the main benefits of using a GIS rather than a normal data base to investigate relationships in your data.

Spatial Query: Another name for Select By Data Layer.

Spheroid: Planet Earth is not a true sphere. Instead, it is more like a sphere that has been squashed at the ends and has bumpy bits (mountains) and troughs (ocean basins and trenches) all over it. Therefore, in order to produce an accurate map, this slightly non-spherical shape has to be accounted for. This is done by producing a model called a spheroid. Different projections and coordinate systems use different spheroids, which are defined by the datum which is associated with it. For the most part, this is not something you need to worry about except to make sure that you use the same datum for all data layers in a GIS project. If you do not, you may find that they do not overlay each other properly.

TIN: See Triangulated Irregular Network.

Triangulated Irregular Network: A triangulated irregular network, or TIN for short, is a continuous surface where triangles of different sizes and shapes are used to define the values for each point within the area covered by the data layer.

Union: To combine data from two or more separate data layers into a single new data layer. In general, the attributes of all data layers will be maintained, but features will be split where they intersect each other. However, in order to 'union' two or more data layers, they must be of the same type (e.g. two polygon data layers, and not a polygon and a line data layer).

Vector Data or **Vector Data Layer:** Another name for feature data or a feature data layer.

Vertex: A vertex is a point on a line or a polygon feature. Usually (but not always) vertices are the points where a line changes direction, or mark the corners of polygons, and their positions are defined by a set of X and Y coordinates. This set of X and Y coordinates tells the GIS software where a specific feature should be plotted. The path of a line or the shape of a polygon can be changed by changing the position of one, or more, of its vertices. In addition, the path of a line or the shape of a polygon can be changed by adding additional vertices, which can then in turn be moved to a specific set of X and Y coordinates.

Vertices: The plural of vertex.

Waypoint: A waypoint is a location in space represented by a set of coordinates which is either part of a survey track or which represents a single point in space where an object of interest is located, a point where something happened or a point where something was recorded. The name comes from the use of GPS receivers which can record positions and store them as a list of waypoints.

X-Y Data: X-Y data are data which are held in a table that also has coordinates which identify where each data point lies in space. Commonly for biological data, these coordinates consist of latitude (the 'Y' data) and longitude (the 'X' data) recorded from a GPS receiver. In addition to the coordinate data, there may also be other information in the table such as an ID number, what was recorded at that position, and other non-spatial information of interest. In general, X-Y data sets are often created in other types of software, such as databases or spreadsheets. They are then imported into a GIS and plotted using the X and Y coordinates to create a point data layer. Once added, such point data layers can be used like any other data layer. An example of how to do this can be found in exercise two in chapter thirteen.

Zonal Statistics: Zonal statistics allows you to summarise information from a data layer based on the zones assigned to them by features or values in another data layer. Information which can be calculated usually includes the mean, or average, the count of data points in a zone and the standard deviation of the cell values within each zone. Zonal statistics are typically applied to raster data layers based on zones in a polygon data layer.

Index

Made in United States
North Haven, CT
25 January 2023

31624349R00193